Risk Management

Recent global developments have fundamentally changed the landscape for conducting businesses and undertaking social transactions. It is more important than ever to step back and take a broader view of managing risks across all areas of the business. Aimed at small- and medium-sized enterprises, which represent the majority of business units globally, *Risk Management: 10 Principles* outlines the ten major risk factors every business must consider when establishing a risk management system.

Now fully updated after being in print for over 20 years, this book focuses on taking a practical and holistic approach to identifying potential risks and considering how best to manage these risks. Modern topics cover the move to home working for many because of COVID-19 and increased risk in supply chains, plus many more. Case studies and self-reflection questions are included throughout, with additional checklists available to download to help you focus on particular areas of the business. This readable and entertaining text intends to help the non-expert establish a workable risk management system across all parts of their organization, especially in smaller business units where a risk management professional is not part of the workforce.

Restructured and reordered to bring the content right up to date, this book is geared towards those working in small- and medium-sized enterprises, business consultants working with client companies, and as an introductory reader for those new to the concepts of risk management, including safety and environmental consultants and students on occupational health and safety and environmental studies courses.

Jacqueline Jeynes is a professional consultant and non-fiction author based in Wales, UK. Jacqueline was Health and Safety Policy chair for Federation of Small Businesses for more than 15 years, a member of UK government organizations including the Health & Safety Executive and British Standards Institute and a regular speaker at international conferences. She has been course writer/tutor with various universities including Durham University (MBA), Aston University (Business degrees), and Aberystwyth University (as distance learning tutor). With 13 books published and many professional journal articles, Jacqueline continues to be in demand as a public speaker.

Risk Management

10 Principles

Second Edition

Jacqueline Jeynes

CRC Press
Taylor & Francis Group
Boca Raton London New York

CRC Press is an imprint of the
Taylor & Francis Group, an **informa** business

Designed cover image: © Shutterstock

Second edition published 2024
by CRC Press
2385 NW Executive Center Drive, Suite 320, Boca Raton FL 33431

and by CRC Press
4 Park Square, Milton Park, Abingdon, Oxon, OX14 4RN

CRC Press is an imprint of Taylor & Francis Group, LLC

© 2024 Jacqueline Jeynes

First edition published by Elsevier 2002

ISBN: 9781032522302 (hbk)
ISBN: 9781032520964 (pbk)
ISBN: 9781003405641 (ebk)

DOI: 10.1201/9781003405641

Typeset in Times
by Newgen Publishing UK

Contents

PART 3   Controlling the Risks

Acknowledgements

This is the revised version of my original publication *Risk Management: 10 Principles* published by Butterworth Heinemann (later Elsevier) in 2002. As it is still being read and copied more than 20 years later, it seemed appropriate to revise and update it, and particularly to make it relevant worldwide.

I would like to thank all those who originally listened to my ideas for the 10 Principles and many people within the Health and Safety community who have been discussing these issues with me for a very long time!

Part 1

Introduction

1 How Can This Book Help Me to Manage Risks?

1.1 AIMS OF THE BOOK AND HOW IT CAN BE USED

As it is now more than 20 years since the original version of *Risk Management: 10 Principles* was published, there have inevitably been significant changes in the way businesses operate and the risks they face. Since 2020, we have seen unprecedented global turmoil and challenges to the way we live and work that could hardly have been anticipated, certainly not on such a scale. The global pandemic was one of the greatest challenges, presenting wide-reaching risks to the movement of people, vehicles and goods, information, and the way business transactions are carried out. Add to this the impact of global conflicts and the very real threat posed by fuel/power shortages, it is now more important than ever to step back and take a broader view of how we manage risks across all areas of the business.

Setting up and running a business of any size is "risky", as is life in general. No situation is risk-free and, of course, not all risks are negative. This book is aimed at small and medium-sized organizations that make up the largest proportion of all business units across the globe, though not necessarily as the largest employer.

It is, therefore, ideal for the non-expert rather than as an academic text, providing a practical starting point for introducing a relevant, workable, risk management system in your own organization. While it is not presented in the same format as an International Organization for Standardization (ISO) Management System Standard, for instance, it is underpinned by these same principles.

1.2 WHAT DOES "RISK" MEAN FOR DIFFERENT TYPES OF BUSINESS STRUCTURE?

There are differences in the way businesses are organized depending on their size and industry sector, as are the type of risks they face. This book is aimed at smaller firms, those with fewer workers and access to Health and Safety or Risk specialists in-house compared with larger organizations that are more likely to have a broader management structure in place.

As we see changes in the way businesses operate globally, earlier definitions of a "small firm" based on the number of employees become less helpful though still a useful indicator. A combination of number of workers/annual turnover/industry

DOI: 10.1201/9781003405641-2

sector/or incorporation status is a better option when we consider potential risks and realistic ways to manage them.

Depending on how you define your business, the following summary highlights the range of issues that are likely to have the greatest impact.

1.2.1   ENTREPRENEUR, SOLE TRADER, AND PARTNERSHIP

Often unincorporated, the owner is the critical person and central point of the organization's culture and so has a fundamental impact on how risks are controlled. A more fluid, flexible management approach to evaluating risks suggests that potentially they can be spotted and dealt with quicker than in a business with a more complex structure.

A significant issue in relation to managing risks is the problem of keeping up-to-date with what the regulations need you to do as the focus is on providing the product, or service, rather than administration. For partnerships, issues relate to control and decision-making in relation to managing risks.

While you may be familiar with the more obvious safety hazards related to your product or service, based on experience, there are often broader risks that are less familiar. This guide will help you to look out for the less common hazards and think about how you can control the risks to safeguard the business and your staff.

1.2.2   MICRO AND SMALL FIRMS

As a small business grows, the culture and beliefs of the original owner will underpin the approach to identifying and managing risks in each organization.

There are many definitions of what a micro or small firm is and the way regional legislation has to be applied. For example, in the United Kingdom (UK) a micro-firm has up to 10 employees and a small firm has up to 50. In European Union (EU) countries, a firm having up to 250 employees is considered a small firm; in the United States (US) the most common form of business unit has fewer than 20 employees, around 89% of all businesses. In Canada, a business unit having fewer than 20 employees is considered a small firm, but even if there is only one individual, there are certain regulations that still have to be complied with.

Given the different regulations that apply globally, this is a further issue to take into account. Given such significant differences in what you are required to do, we cannot provide an over-arching "must do" list that will fit everyone. We can, however, highlight the risk factors that could apply to your business so that you can make an informed judgement about how best to manage them.

1.2.3   LARGER ORGANIZATIONS

Our focus is on smaller firms, but you may be part of a small business unit that has to deal with specific risks outside the control of a parent company. This does, of course, bring its own problems related to authority, decision-making, and cross-functional communication. However, you should still be able to gain some useful insights by working through this guide.

Although not exhaustive, the above summaries highlight some of the potential risk factors that firms must consider irrespective of size, and the need for a systematic approach that is relevant, comprehensive, and cross-functional while acknowledging the unique spread of pressures facing each individual firm.

1.3 THE IMPORTANCE OF PEOPLE IN MANAGING RISKS

We will return to the "principles" behind risk management later, so the focus in this section is on the major risk factors facing any business.

In a small firm, people within the organization are a crucial element in the way risk is identified and controlled. Everyone brings their own perception of what they think of as risk while in the workplace, often related to previous experience. There is often reference to "common sense", and therefore an assumption that a condition is obvious but, of course, this comes back to an individual's knowledge and understanding.

While this represents internal factors that you can manage, there are inevitably conditions that are outside your scope of control. Figure 1.1 identifies the main external pressures you are likely to meet, although there are likely to be others that relate specifically to your industry sector that you will already be familiar with. When considered alongside the changing and uncertain face of current competitive climate, we can see why risk management is often sidelined in smaller organizations!

As you can see, these external factors are still affected by how people throughout the organization see their role in ensuring that potential risks are identified, and how they then ensure controls that are put in place are sustained.

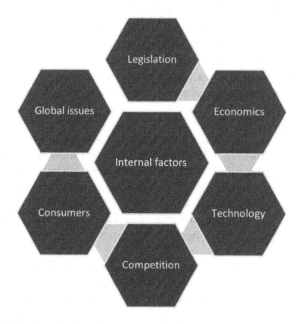

FIGURE 1.1 Internal and external pressures on business.

1.4 THE 10 PRINCIPLES OF RISK MANAGEMENT AND RISK FACTORS FACING BUSINESSES

The risk management approach identified by the author in *Practical H&S Management for Small Businesses* (Jeynes J 2000) was developed specifically to help those working in or with small and micro-firms. It considers the health, safety, fire, and other legislative risks to the business. Primarily concerned with physical or visible risks, so making it easier for a non-specialist to establish a workable management system, it is a useful starting point for developing a more holistic system to analyse and evaluate business risks.

By adding further elements of planning and performance measurement it becomes a much more comprehensive management tool that can be tailor-made to suit the individual firm. We have used the 10 Principles to focus on the risk factors facing business. The principles behind this approach are more about the actions needed to fully engage in managing risks in your business.

The risk management approach in this guide is therefore based on the following actions:

- Take a holistic view of risks to the business covering all functional areas rather than individual elements that never seem to give the full picture.
- Consider the same potential risk factors against each function of the business, even if their impact is considered minor in some areas.
- Include input from all levels of staff within the organization, with commitment at the very highest levels.
- Consider potential employment risks for the present and the future.
- Consider potential technical and production risks.
- Consider health, safety, environment, and security risks.
- Consider potential legislative risks.
- Consider economic risks.
- Include consideration of trade and business risks.
- Ensure policy and strategic planning for the future reflect the results of assessing risks in all of these areas.

There is never a "quick fix" that can be applied to and reflect every size or type of business structure. However, the ten principles and ten factors will help to cover the main areas of establishing a workable risk management system.

An evaluation of all the business operations requires honesty and motivation if you wish to produce a comprehensive, detailed analysis of potential risks across all areas. It is a daunting task, but a necessary one to gain a true appreciation of how all the elements fit together, rather than the "sticking plaster" approach to dealing with risks piecemeal as they materialize.

Figure 1.2 illustrates the links between the different factors and how they impact each other. Although the ten principles cover the main risk factors comprehensively, it is hardly a nice easy number to remember! They have, therefore, been broken down into four distinct groups:

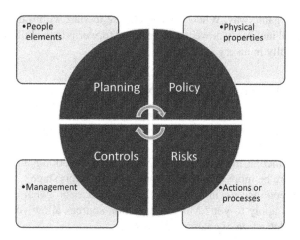

FIGURE 1.2   Links between the risk factors.

1. **Physical properties – Premises/Product/Purchasing**
   Premises – logistics, transport routes, business travel, sales channels
   Product – sourcing materials, climate change, and protecting the environment during manufacture and disposal, waste management, new designs
   Purchasing – pricing, delivery, legislative barriers and tariffs, global supply risks
2. **People elements – People/Performance/Protection**
   People – attitudes to risk, skills, training, and worker legislation
   Performance – quality and environmental controls, marketing, sales, incentives, competitor risks, finance risks, and stakeholder expectations
   Protection – including security risks for data, premises, people, patents, and designs
3. **Actions or processes – Processes/Procedures**
   Processes – risks to workers and users, technological developments, where and how staff work
   Procedures – relevant and up-to-date, able to deal quickly when new risks appear
4. **Management issues –Planning/Policy and Strategy**
   Planning – involving others across all functional areas, gaining commitment, recognizing, and planning for future potential risks
   Policy and Strategy – realistic and workable, based on identifying and managing all risks

These all overlap or interact with each other constantly, so you cannot easily separate them. However, they do provide a structure to help you identify and evaluate risks to the business, and to initiate/monitor controls to reduce these risks.

Many guides focus on the policy of the firm as a starting point although, as we know, businesses tend to start from the concrete and move on to strategic issues in a more untidy, organic way. Wherever you choose to start within the **10 Principles** structure, you will inevitably move backwards and forwards to the Policy and Planning elements, especially in a very small firm.

1.5 SELF-REFLECTION QUESTIONS

- How would you define the structure of your business as it broadly fits with the descriptions outlined above (points 1–3)? An organization chart might be a useful aid here.
- Is there someone in your business who will be working through the book? Should there be more than one person involved at this stage?
- How do you access information and guidance on what local rules and regulations apply to you? Most information sources allow you to sign up for regular updates free of charge.
- Having collected all existing risk assessment documentation, can you identify where the gaps are to ensure you are not spending time duplicating something that is already available?

Part 2

---

*Assessing Risk Factors in
Your Business*

2 Identifying Risk Factors and Assessing the Risks

2.1 DEFINING RISK ASSESSMENT AND RISK MANAGEMENT AND DIFFERENT VIEWS ABOUT RISK

Definitions of risk management vary depending on the underlying purpose of why/how you carry them out and your specific focus for each of the activities. There are similarities between definitions, so we will focus on elements that are of most importance to a small firm.

While many risk management definitions focus on financial risks, as in the Risk Management Framework (RMF) of the US, this is not our primary focus here. Inherent in the 10 Principles approach is the combination of business continuity and non-business risks, such as risks to the people who work there, as they are so closely linked in a smaller organization.

2.1.1 DIFFERENT APPROACHES TO RISK MANAGEMENT

Often used as an overall term, there are different ways to classify what we mean by "risk management". Taking the simplest approach (the four Cs (control/communication/capability/culture) of risk management), the four steps will be:

- Identification of hazards
- Analysis of potential risks
- Evaluation of controls in place
- Treatment to reduce risks.

In this format, the first three stages are about risk assessment, and the fourth stage is about risk management. A wider definition as six components (Worksafe in N.T. Australia) includes the first three stages as risk assessment and the final three stages as risk management:

- Hazard identification
- Risk identification
- Risk assessment
- Risk control

DOI: 10.1201/9781003405641-4

- Documenting the process
- Monitor and review.

If we broaden the definitions to the overall process of risk management, the ISO 31000 outlines 7 Rs of risk management (ISO Consultants) as follows:

- Recognition
- Ranking/evaluation
- Responding
- Resourcing controls
- Reaction planning
- Reporting and monitoring performance
- Reviewing risk management.

When we come to the issue of risk control, there are four basic options: avoidance/reduction/transfer/retention (Unit4.com). Perhaps five options with a little more detail – avoidance/retention/spreading the risk/loss prevention, and reduction/ transfer through insurance and contracts.

As health and safety legislation around the world depends on a risk assessment approach to managing and controlling hazards, there is a lot of information and guidance available on what this involves. While this book is not exclusively concerned with health and safety risks, there is a legal requirement to conduct such assessments; therefore, it is prudent to start from this point.

The principles of risk assessment are quite straightforward, based on the following activities.

1. Identify hazardous conditions/properties/processes that could potentially cause harm, injury, or damage.
2. Consider what this harm, injury, or damage might be; who could be affected; and how serious the result of exposure might be.
3. Evaluate the likelihood that such harm, injury, or damage will occur, taking into account any control measures that already exist.
4. Make judgements about how adequate these controls are, identify gaps where you think they are not good enough, and prioritize actions needed to correct the situation.
5. Monitor and re-evaluate your findings regularly and when circumstances/ materials/processes change.

Despite the simple logic of this approach, there has been much hype and confusion generated about what RISK ASSESSMENT is with the result that the potential value and practical application of the process has become lost under a mountain of paperwork. This does not mean that records of the process are unnecessary. Clearly such activities need to be recorded in some form to confirm they have been carried out adequately, and to ensure that:

a) The scope of activities to be assessed was clearly identified beforehand.
b) The full range of potential hazards or risk factors has been considered.
c) People in the organization know what these are, what controls are in place, and how to use them.
d) Adequate monitoring and review can take place.
e) Other parties can see that risks are being managed appropriately.

Wherever the business is located, the underlying principles are the same – significant findings of health and safety risk assessments must be recorded by law. For example, this applies in Canada if there is just one person in the business; in UK and EU countries it is required if five or more people are employed. So, risk assessments must be relevant to the business, carried out by suitably experienced, competent people, be of sufficient depth to ensure people and property are protected, and reflect what really happens in the firm rather than what management thinks should be happening. This does, therefore, require input from several sources, both users of the systems and managers, technical experts or health and safety specialists.

In the author's view, the least successful approach is the use of external specialists to "do risk assessments" for the firm with little or no input from internal staff. There may well be valid reasons for using external consultants for some elements of the risk assessment, as we can see later, but this should be identified through earlier assessments within the firm. Even the specialist expertise required for assessments such as noise levels, fibre/air concentrations, or asbestos materials will need to involve workers directly.

For some readers, this will be a familiar ground and Health & Safety Risk Assessment will have been a regular occurrence as it has been a legal requirement for decades. Other readers may be familiar with evaluating broader risks to the business but not specifically health and safety.

The intention in this section is to establish a risk assessment approach that can be applied to all the elements of the 10 Principles, bringing together the fire, health, and safety risks with other issues and concerns that need to be controlled and managed in the organization. The starting point in Section 2.2 is to identify hazards within the firm, then evaluate the hazards or risk factors in Chapter 3, and finally allocate risk ratings in Chapter 4.

2.2 IDENTIFYING POTENTIAL HAZARDS

All industry sectors have hazards or risk factors that are so glaringly obvious to those working with them that they are hardly worthy of note. Unfortunately, the frequent reference to "common sense" belies the previous knowledge, skills, and experience of individuals that help them recognize such hazards, and make it less likely that newcomers to the industry are made aware of them.

Example: A young man was killed within hours of starting work at a firm based in the Docks in the UK. Sent there by an employment agency, he was given no training in how to carry out the tasks set, which were particularly hazardous given his position in the cargo hold in relation to the crane grab being used. He was disadvantaged still further by the total disregard for any safe working procedures.

In larger organizations, different sections or divisions may well have a base of knowledge of hazards but no central source of reference. Results can be patchy, depending on the way data has been collected and the range of people involved in its collection. Crucially, knowledge can be held by an individual as a long-term employee, and potentially lost when they leave the organization.

It is important that a risk analysis structure is defined and applied throughout the firm to make it both consistent and easier to collate the findings. This also helps to ensure that all the relevant risk factors have been identified. Records should also include details about where and when the assessment is carried out, and by whom.

A valuable starting point is a site plan, irrespective of the size of the firm, as it gives visual reminders about areas of activity that are sometimes forgotten, such as external waste storage areas and rarely used storerooms.

It is useful for highlighting movement of people and goods through the firm, and potential areas where there are conflicting priorities of use. From an insurance point of view, it also demonstrates that a comprehensive analysis has been carried out, and that the Planning and Policy elements are based on relevant information.

There is a checklist to help identify movement of goods through the business (Appendix Checklist 2.3) if you prefer to use it as a reminder of how/where goods flow. It provides a useful structure to work with, including elements of delivery of materials as well as onward travel, but any format that includes these elements will help to establish a structured approach.

Figures 2.1 and 2.2 illustrate how you might draw up a plan for your site. However, note that a hand-drawn version works just as well! It can work with a service type of business as well as a production facility, with movement through the business of people, animals, or objects. The list of stages can be amended to reflect the work of your own business unit and should reflect what usually happens on a day-to-day basis. It also helps people involved in these different stages to quickly confirm the type and range of activities carried out.

The areas to look closely at include:

- Car parks, delivery areas, and pedestrian access
- Reception and waiting areas
- Storage of supplies and materials
- Movement of goods from stores to first process stage
- Office and administration areas
- Public display areas
- Work in progress stages
- Maintenance and repair
- Activities carried out off-site, including movement between sites
- Storage of waste/scrap materials

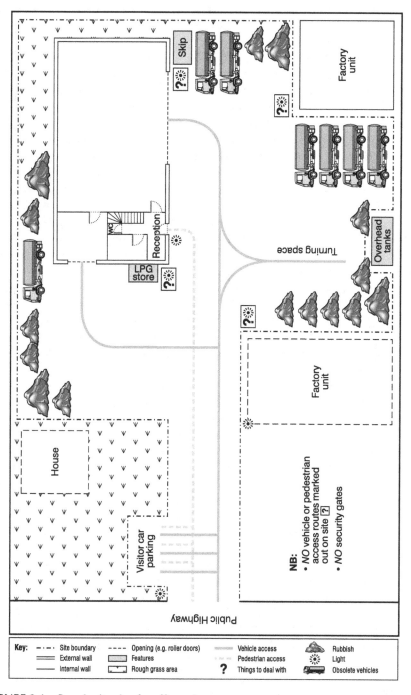

FIGURE 2.1   Sample site plan for office suite.

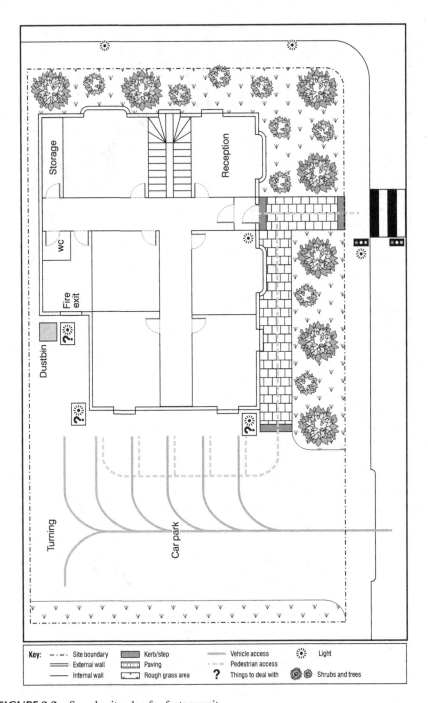

FIGURE 2.2 Sample site plan for factory unit.

- Packaging and storage before distributed
- Distribution.

It is also valuable to consider what equipment or processes are used at each of these stages, especially for health, safety, and security risk factors considered later, and the number of people involved (including those who only occasionally need to be present).

The following section considers each of the ten risk factors identified in Chapter 1, within the groups of

a) Physical properties
b) People elements
c) Actions or processes
d) Management issues.

Following a similar process to that in the three areas of physical properties, people elements, and actions or processes, to identify potential hazards, we can see that the management elements are broader. In reality, they cut across all factors and functions of running a business, but we can still look at them through the same five headings.

This will lead to thinking about policy development later in our risk management process – see Chapter 7 in Part 4. For now, we have just listed the areas where you will need to have a policy statement in place.

2.3   RISK FACTORS

2.3.1   PREMISES/PRODUCT/PURCHASING

This is often a significant group of risk factors for smaller firms, as they frequently have limited access to suitable premises either at start-up stage or when expanding production. A wide range of risk factors include suitability and the size of premises, financial concerns related to ownership or tenancy status, and location.

Several risk factors are associated with the product/service, such as product life-cycle, life-style trends and demographic changes, your competitive position now and in the future, and, of course, environmental protection issues.

Purchasing is often isolated from consideration of other elements, but should refer to the type and availability of materials, technology, and plant capabilities. Broader issues of external standards are also important.

As regulations become more complex and far-reaching, there is sometimes more scope for conflicting requirements from different enforcement bodies on the same issues, such as health and safety, fire, and planning requirements for your premises.

Changes outside your control, such as those related to accessibility and fire precautions regulations, may require significant alterations to premises. The age, type, and structure of the buildings will be a major factor, of course, but so too will be the ownership status or responsibilities of the occupier. Multiple occupancy premises may have substantial difficulties in this area.

We have seen society generally becoming more litigious, so potential hazards associated with the product or service need to be carefully considered. Providers of information, guidance, and advice are just as susceptible to such hazards as producers of consumer products.

While a wide range of regulations have been introduced for transporting hazardous goods or materials, particularly across borders, there are often specific systems and qualifications required at all stages of the journey whatever the product or service.

Many firms do not review purchasing arrangements to consider the optimum option but stay with the same suppliers for many years. While such loyalty sometimes results in preferred supply conditions, the risk factors associated with this approach are varied. Generally, these risks relate to being able to change supplies easily to match changing processes or methods of production.

Substitution of hazardous substances in printing, for instance, may represent competitive risk factors if sufficiently robust or effective substitutes are still not available. For example, the use of water-based inks for printing on flexible plastic products results in print that does not always "fix" satisfactorily.

Some countries have banned the production/sale of plastic bags and packaging, especially single-use products. Packaging that is robust but also seen to be eco-friendly and biodegradable is likely to become more of an issue in the near future.

Employment

General ageing of the working population means fewer young people in some industry sectors, such as nursing or care sector where the most experienced and skilled groups of workers are expected to retire during the next decade. More workers are now caring for older relatives, as life expectancy increases, as well as children or young adults who are living in the parental home for longer.

The following list identifies potential issues related to employment:

- Location – skill shortage areas and access to suitably qualified or experienced staff, particularly in areas of traditional industries now redundant, or where recruitment of overseas expertise is needed
- Capacity of premises for workers and customers; facilities for staff and customers
- Capacity for production processes and people to operate them
- Effective communication between all parts of the firm, especially if based in different locations

- Direct contact with the public, high levels of stress; potential/actual violence to workers
- Insufficient input from all levels within the firm; exclusion of some groups of workers, especially those working on shifts/off site/travelling around from site to site
- Balance between needs of different sites or business units.

In industries such as agriculture and horticulture, it is sometimes difficult to receive supplies in size of container and quantity that reduces risks of manual handling injuries without significantly increasing the cost of materials. In addition, agriculture and horticulture, scientific testing facilities, food production, and medical services are all sectors that need to identify potential risks to health from exposure to biological or chemical substances.

Legislation

Older premises are not easily converted to reduce potential safety hazards of narrow or uneven stairs and steps. Use of asbestos materials in the structure of buildings needs to be identified, potential risk to workers and others need to be assessed, and removal, where necessary, needs to be carried out under controlled conditions.

Increased pressure to provide smoke-free working environments may not present problems where spare capacity is available for specified smoking areas.

In addition, environmental and transport policies produced by local authorities increasingly include restricted access to town centres at certain times of the day with specified times for loading or unloading vehicles, charges, or tolls.

Consider the impact of the following legal requirements:

- Building safety – structure of buildings including narrow or uneven stairs and steps; layout of buildings for moving goods or materials
- Product safety – potential for injury if not used properly; labelling and packaging; sharp edges and/or moving parts; weight/size/shape of product; child safety requirements
- Worker safety – changes in law regarding use of materials/substances; greater specification of safe methods of working in some industries
- Site safety – adequate storage facilities; safe loading/unloading areas; use of vehicles on site; storage and handling of hazardous substances and materials
- Electrical supplies; adequate lighting and potential for slips and trips; working areas considered to be "confined spaces"

- Health of workers – heating and lighting provision; ventilation and clean air; smoking of workers or customers; washing and toilet facilities; extreme heat/cold areas in workplace
- Identify and record all areas of the site where asbestos materials may be present; damaged or exposed sections of asbestos materials; potentially hidden sources of asbestos, particularly older premises in the UK and EU countries
- Fire spread – structure of buildings including open-plan areas and stair-wells; access routes; escape routes/facilities available in emergencies; structural materials; location to other high-risk premises; shared premises and multiple occupancy
- Fire protection – protection of surrounding areas in the event of fire; impact on people/buildings/land/animals
- Environmental protection – disposal and storage areas for hazardous materials; proximity to residential areas regarding noise emissions; ventilation and exhaust systems used
- Energy use – economic use of power sources; use of renewable sources
- Transport – transporting goods or substances; qualifications of relevant staff
- Permits or licenses to operate in given premises; license requirements to safeguard workers and public; planning permissions needed, particularly if premises are based in different countries/regions
- Competition laws and the need to operate a fair procurement system
- Replacing plant and machinery with safer/more environmentally friendly/more efficient/quieter versions
- Exposure to violence and/or stress; working from home; working in other people's premises
- Emergency procedures in the event of a fire, explosion, spillage, or leakage of substances
- Conflict between different enforcement bodies on acceptable actions required to comply with the law
- Changes to law that require alteration to premises; problems associated with age/type of premises and ownership status.

Some industries find it difficult to recruit relevant skilled and qualified staff that reflects the make-up of the local community, so leaving themselves open to charges of non-compliance with relevant inclusivity or anti-discrimination legislation.

Security

Access to the site may or may not present additional risks, but unauthorized entry to shop floor areas by delivery staff, for instance, may introduce further hazards.

Potential for arson can be difficult to identify in some situations. Piles of broken pallets against buildings near to perimeter fences are ideal targets for arsonists, as are waste skips full of flammable materials. Less obvious targets may be closed or "secure" storage areas that require more effort and planning on the part of the arsonist.

Potential security issues are likely to include the following:

- Visitors to the site and control of access to certain areas; perimeter fencing and gates
- Unauthorized visitors – potential for injury; arson; property damage; vandalism
- Theft of materials; goods produced; workers'/customers' valuables; safe storage areas
- Access to electronic data; personal records; confidential information; growth in identity theft, sale of data to outside agencies and potential for criminal activities, cyber-crime
- Access to harmful substances/agents; storage of hazardous substances and materials
- Violence to staff, visitors, and animals on site.
- Transport safety and security of vehicles and personnel in transit between sites; movement of goods/materials between sites; use of refrigerated or other specialist vehicles and containers; movement and storage of money and valuables
- Confidentiality of personnel assessment reports and feedback
- Breach of confidentiality on company-wide or divisional performance inside and outside the organization; leaks of information to markets or media.

Often overlooked, the security of vehicles is a risk factor whether they are stored on site or used by staff to travel to and from work. Delivery of materials, components, and finished goods should be considered, as well as transport of equipment or plant to carry out the job off-site (such as construction).

Competition

Copying high-profile branded goods is a multi-million-pound business world-wide, especially fashion items and sports gear. Protecting brands and designs may therefore represent a considerable risk factor.

Competitive issues to consider include the following:

- Copyright, patents, registered designs; industrial espionage (intentional or unwitting)
- Benchmarking or third-party initiatives with potential competitors in same industry
- Declining or growing commercial locations – impact of nearby derelict or empty premises on customer perception; customer expectations of premises, image, facilities
- High-street or out-of-town location
- Demographic changes in customer profiles relative to product or service
- Changes to legislation regarding use of materials/products across geographical borders
- Ethical approach; environmental protection; use of environmentally friendly materials; public perception; major move to eliminate the use of plastic carrier bags generally in retail (even if new biodegradable materials are developed)
- Public perception of industry and firm; bad publicity for other businesses in industry
- Comparative costs of production when finding new trading partners or renegotiating existing trade deals
- Changes to local by-laws that make deliveries more difficult or time-consuming – local authority changes to parking and loading facilities; changed one-way road systems; diverted road and pedestrian traffic
- Use of outmoded, slower methods of production; outmoded processes and falling behind competitors
- Need to replace plant and equipment with more efficient models; ability to meet customer demand for new features or facilities of product or service
- Customer service expectations; replacement/refund/compensation procedures
- Undue reliance on limited number of suppliers of components
- Transport and delivery costs; infrastructure weaknesses in region/country-wide
- Product life-cycle stages; access to research and development facilities.

Substitution of hazardous substances in printing, for instance, may represent competitive risk factors if sufficiently robust or effective substitutes are still not available. For example, the use of water-based inks for printing on flexible plastic products results in print that does not always "fix" satisfactorily. Some countries are already banning the production/sale of plastic bags.

Finance

Risks associated with mismanagement of the purchasing or procurement function may be significant. The larger the organization, the more potential for problems related to communication channels and inappropriate ordering schedules. This may be exacerbated by central purchasing policy that does not take full account of individual business unit requirements in other countries or regions.

Consider financial implications in the following areas:

- Cost of materials, duties, legal costs; changes to local authority and utilities charges
- Terms of leasing agreements; timespan for agreement renewal
- Life span of plant; mark down and write-off values; cost of investments; repair and replacement of equipment/machinery/buildings
- Maintenance requirements; shared premises and responsibilities
- Insurance premium rates dependent on growth or decline of local environment
- Additional security precautions required due to location, such as CCTV or metal shutters
- Profit ratios; costs of production rising/steady/falling
- Range of suppliers available; payment or delivery systems
- Storage and warehouse facilities and costs; call-off facilities; use of Just in Time for materials or components; stock control; penalties
- Delivery methods and costs; cost of fuel; vehicle and other license costs
- Collection of payments; debt/credit control systems; electronic transfer of payments; payroll systems
- Cash flow; banking facilities; lending rates; expected return on investments
- Value added tax and other tax regimes for firm's products or service; base rate/interest rates for borrowing/tax rates and incentives
- License costs; worker training and qualification costs
- Opportunity cost of holding large quantities or high-value stock
- Cost of implementing and maintaining third-party certification schemes
- Short- versus long-term investment programmes; potential returns on investments; investment time scales for rapidly changing technology; access to affordable finance for investment
- Withdrawal of financial backing; loss of trading value on world stock markets; higher cost of lending; exchange rates/strengths and weaknesses of world currencies
- Government actions in future that impact long-term planning programme
- Consumer spending trends at home and abroad; inflation; regional upturns or collapses
- Uncontrollable events external to firm – for example, fuel price rises or shortages.

Serious flooding in many parts of the world previously unaffected has already resulted in massive increases in insurance premiums, limited insurance cover, or no cover at all for future flood damage. The situation is similar globally – for example, in the UK during 2000, 2015, and 2017 and Texas in 2017 – inevitably resulting in additional local authority bills as they seek to recover costs of clearing up and improving flood defences.

Uninsured losses can be up to ten times more than insured losses. Insurance protection generally only covers around 20% of the real cost of incidents, and indeed many people are already under-insured. Operational changes may not have been notified to insurance provider for some time, and values for plant may be much higher than original premiums allowed for.

2.3.2   People/Performance/Protection

It is important to consider workers at all levels in the firm, especially those with non-traditional forms of work contract and temporary workers. Risks to visitors to the site, and wider public in the vicinity, should be identified.

Issues will include the culture in the workplace, how workers are organized, and skills and competence of workers.

Performance can be viewed at individual/department/company level, either as part of a benchmarking exercise or just related to a particular firm. This relates to performance measures, and who wants to see them.

Protection is much broader than protecting people from health and safety risks and includes identifying risks associated with premises, materials, intellectual rights, data, and security.

The scope and range of disciplinary issues stated in employee handbooks sometimes become so great as to be a charter for dismissing as many people as possible in the shortest space of time! Assuming that employees are necessary for the firm to operate, performance measures should optimize performance to help the firm achieve its objectives, rather than as a route to inevitable failure.

Historical data on accident rates per annum or sickness absence may be patchy or incomplete. Any data collected needs to be analysed closely to provide indicators for future targets. Investigating accidents is an area often neglected in firms, though a legal requirement, especially those with a blame culture that results in punishment for anyone recording accidental damage to people or property.

Performance measures must reflect what people do to complete the task, and therefore must have input from workers themselves. Where people have negative experiences of trying to reach unrealistic performance measures, they tend to be more sceptical of others that are introduced. Good communication skills are vital.

Employment

Loss of major industries such as mining, heavy manufacturing, or banking represents significant risk factors to firms in the area. The base of highly skilled labour from these industries often needs substantial retraining for work in different sectors.

Training is a crucial element of maintaining the skill base, and the type of skills training required by the firm may not easily be accessed locally. Risks are also associated with the format or structure of available training, or the ability to analyse exactly what is needed. However, the rise in online training modules has benefitted workers and employers.

High staff turnover and use of a large proportion of workers on temporary contract represents an additional risk. This is more apparent when linked to the need for providing adequate health and safety training for staff.

The following list identifies potential issues related to employment:

- Age structure of workforce and demographic trends
- Range of skills and experience of current workforce and match with activities
- Willingness and ability of workforce to upgrade their skills; skill levels and experience of sales force, purchasing team, workforce in the future
- Need for flexible staffing; increased workloads and insufficient staff numbers; too many people and not enough work to gainfully occupy them; fluctuations in customer demand/seasonal work; use of temporary staff
- Need to find relevant training provision; internal or external provision; training and support available internally; cost and time involved, motivation and commitment of staff
- Amount, type, quality of supervision, and management of workers; organization of work groups and communication links between them
- Culture of organization; review and feedback mechanisms available
- Increased use of technology to reduce the number of people employed per business unit; redeployment of staff and resources; recognition of union membership
- Opportunities for promotion and development of staff; barriers to advancement; facility for staff to move between departments
- Inexperience of workers; high proportion of newcomers to industry, for instance, from redundant industries in the region; significant proportion of "vulnerable" groups of workers, especially young people
- Relationship with discipline and grievance procedures; are they used as a positive rather than a negative management tool?
- Violence to staff from customers/other staff/unauthorized visitors.

Legislation

Fire safety regulations have shifted to emphasis on a risk assessment approach, with particular reference to the potential impact on, and protection of, the local environments if a fire occurs.

Emergency drills for dealing with a fire, restricting its spread, and evacuating people can only be tested for validity in the event of a real fire. Many people do not take fire drills seriously, and often do not know what the procedure is in their workplace. For instance, few bars or restaurants ever hold a practice evacuation of the premises even though their customers may be very vulnerable in the event of a fire.

As worker protection measures increase, so too do risk factors associated with balancing these measures against the need for a flexible workforce and the needs of different groups of employees on different employment contracts. High turnover of staff due to unsatisfactory work conditions or requirements is inevitably a high-level risk.

Whether the firm employs many or few workers, they are a crucial element of providing the product or service to the customer. Retaining a committed workforce is, therefore, vital, so organization of the process should support this. The volume of work and the way workers are organized are significant factors, as are access to relevant information and input to decisions that affect them directly.

Consider the impact of the following legal requirements:

- Growing base of employment protection and other government-led measures represent burdens on firms; worker protection measures are legally required
- Use of different employment contracts; conflict of definition between worker and self-employed person; increase in zero hours contracts, flexible working, and part-time requirements; holiday and break entitlements
- Equal opportunities and other discrimination laws relative to potential workforce base and accessibility; reflection of balance in local community
- Employment law requirements related to dismissal, redundancy, time off, contract of employment, discipline, and grievance procedures
- Provision of state-defined benefits, maternity and parental leave, for example
- Safety – adequate training and supervision; provision of relevant information; use of adequate protective gear/equipment; safe equipment, machinery, materials; regular maintenance procedures; driving safety on behalf of firm; lone working
- Health – health screening and surveillance required; lighting, heating, ventilation; use of visual display units (VDUs); sight and hearing testing may be necessary; exposure to hazardous substances; potential air contamination and biological hazard exposure; extremes of temperature
- Amount of record-keeping and monitoring activities required by law

Security

Collection and analysis of data from different sources or divisions within the organization may be time consuming and fraught with problems, although this will inevitably be reduced with the growing use of electronic means for transferring data. Security of this data inside and outside the firm may pose additional risks, particularly if speculative ideas are being formulated in the early stages.

Potential security issues are likely to include the following:

- Time limits on copyright, registered designs, and patents
- Research and development security; technological use in production or materials used for product/service; use of temporary workers; fraud; industrial espionage potential
- Storage of high-value components/work-in-progress/finished goods
- Protection of money and other valuables; amount of money on premises at any time
- Checking supplies in to confirm quantity and quality of goods, checking materials out
- Fire; damage; vandalism; arson; pilfering and theft by staff/others
- Safe storage of hazardous substances; authorized access
- Emergency procedures in the event of theft, violence, damage, and arson
- Systems to monitor movements of staff, particularly lone or mobile workers.

Competition

Firms with stock market listings can suffer serious damage from the lack of adequate security on information. Even smaller organizations not listed can suffer damage from leaks of sensitive information about their plans for future development to the local media.

Loss of credibility or positive image with customers from actions by others in the industry is a potential risk. This may relate to methods of selling or to the product/ service itself such as incorrect advice to clients on pensions during the 1980s and 1990s.

Competitive issues to consider include the following:

- Generation and protection of ideas by workforce; opportunities for workers to put forward ideas and take part in consultation with management
- Consumer protection – use of non-hazardous materials; description of product or service and customer perception; negative feedback posted on social media platforms with immediate impact

- Competition laws and the need to operate a fair procurement system
- Life cycle stage of product or service; consumer trends and cyclical demand
- Target quality levels; quality control and assurance methods
- Introduction of new legislative requirements; pressure for move to a cashless society
- "Poaching" trained staff by other firms; recruitment strategies
- Use of third-party certification schemes for Quality/Environment/ Occupational Health & Safety management systems (such as ISO 9000/ISO 14001/OHSAS 18000 standards); suppliers or customers introducing different management or control systems; need to update procedures and reporting systems

Third-party certification schemes can offer valuable "badging" opportunities for firms, but inability to maintain the standards set or withdrawal of the certification is a risk factor to be considered.

For firms without third-party certification systems in place, these may represent a risk factor if clients use them as a means of restricting access to goods or services. If you are a user of such schemes, there are several risk factors associated with their use, including time and human resources to establish and maintain the system/cost, often becoming a net cost to the firm/difficulty in integrating two or three different standards for Quality, Environment, and Occupational Health and Safety (OH&S).

Finance

The cost of accidents is underestimated in firms that do not have direct experience of incidents at the workplace. While direct costs may be obvious, indirect costs related to loss of production time, impact on witnesses, management time to investigate accidents, and impact of bad publicity locally are often overlooked.

Consider financial implications in the following areas:

- Pension provision; other benefits and incentives; profit sharing schemes
- Minimum or living wage (in the UK) and pay structures, ratio of wage bill to sales
- Pricing strategy and sales methods in operation; access to appropriate materials and suppliers; payment systems and issue of late payment
- Cost of providing a wide range of protective measures; need to keep up-to-date with changing conditions; cost of accidents and injuries if poor safety performance

- Resources required to maintain systems and thus taken away from primary production
- Insurance premiums; costs of injury and ill health to workers; provision of occupational health protection schemes; potential provision and cost of rehabilitation services
- Sickness absence; cover for other periods of absence; maternity benefit provision
- Cost of training and retraining staff; cost of recruitment.

2.3.3 PROCESS AND PROCEDURES

Risks associated with the process itself vary depending on the type of business, but there are fundamental questions related to techniques and inherent risks, controls in place, potential impact of technological developments or changes in legislation.

The hazards relate to how procedures impact on other factors, particularly the Product, Process, and People. The sort of questions you might consider should include how appropriate they are for current and future production, are they monitored regularly, do they reduce risks or pose additional ones?

Greater reliance on computerized systems has generated different hazards for workers, with health issues becoming more prominent. Examples include constant use of equipment at call centres, or long periods of time at visual display screens. Given these changes, access to occupational health services is a growing requirement globally and is likely to become a mandatory requirement for most workers in the future.

Traceability of substances into the local environment is a priority for various enforcement bodies, so must be considered a potential risk factor of production processes. Consideration should include deliberate disposal, such as waste products from food production, as well as potential for accidental spillage or leakage of chemicals or biological agents.

Employment

Sectors with high staff turnover rates, such as leisure and catering industries, often have a high proportion of young workers. Potentially, such workers are more vulnerable due to the lack of experience and skills and have not had time to learn a job fully before moving on. However, mature workers with high-level skills from old traditional industries may also be more vulnerable when having to change career direction.

All industries can experience complacency amongst staff where procedures have been in place and remained unchanged for a long time. In production areas, this may lead to "short cuts" to speed up the task, for instance, by overriding safety mechanisms. This may also relate to inaccurate record-keeping and reporting procedures that then represent a risk to the validity of data used to support management decisions.

Violence or abuse of staff by members of the public has grown considerably over recent years and must be acknowledged as a risk factor by any organization that deals directly with the public. Industries where face-to-face contact with the public is involved, crime rates and instances of threatened or actual violence to staff have increased in recent years, posing additional threats to personal security and safety.

The following list identifies potential issues related to employment:

- Skills and expertise required for future work and activities; gap between these two
- Complacency in the use of long-established procedures; development of "short cuts" to inappropriate procedures
- Lack of skills; lack of motivation or commitment; repetitive, low-skill work
- Little opportunity to take control or make decisions about process used
- Inadequate information, training, and supervision
- Team or group working; piece work; targets for production
- Are performance measures appropriate for the range/type of work undertaken; are they relevant, meaningful, and realistically represent measurable targets
- People are involved in setting performance targets for themselves or the team; everyone knows what they are.
- People are trained in the use of performance measures; feedback is provided when and where it should be
- Lack of knowledge or understanding of issues at operational level by senior management
- Health – exposure to stressors; workloads; noise generation and protection for workers
- Ensure changes to procedures do not introduce additional hazards for workers or others.

Legislation

Legislation changes very quickly, so it is vital for businesses to keep up-to-date with changes. This involves time and cost and takes away effort from the primary function of providing a product or service. Access to information in a relevant format is important for all firms.

As society generally becomes more litigious, so potential hazards associated with the product or service need to be carefully considered. Providers of information, guidance, and advice are just as susceptible to such hazards as producers of consumer products.

The following list identifies potential issues related to employment:

- Lack of knowledge and awareness of current legislative requirements
- Worker safety – use of substances and materials; sharp edges and moving parts of product; personal safety when working with the public or animals and potential for violence; specification of safe methods of working in some industries
- Health – biological contaminants during production; fumes when in production or during storage; allergenic or carcinogenic properties; air quality and emissions at different stages of the process
- Fire – product as a source of ignition/fuel/oxygen when in production, use or storage; flame retardant properties; testing requirements; potential for explosion; flammable qualities
- Environment – methods of distribution if hazardous; disposal of product; obsolescence; substitution of materials for production or packaging
- License requirements for provision of service/production/movement and distribution/disposal of waste and scrap materials; specified qualifications for workers
- Health – manual handling of loads; use of vehicles for handling; noise levels; light, heat, and ventilation in storage areas
- Need to analyse procedures for risks to individuals, including safe working practices and permits to work; monitor, supervise, and control use of procedures on a day-to-day basis
- Record-keeping procedures; number of people involved in collecting and collating data; reliance on historical data about instances, relevance and accuracy of data collected; internal and external reporting procedures such as reporting accident, illness, or near-misses
- Deciding priorities for action that acknowledge the legal requirements of business.

Safety hazards must be identified for individuals where necessary. For example, in the short term there may be additional risk factors for pregnant or nursing women, or someone recovering from an accident or illness. In the long term, the additional hazards present for individual left-handed workers operating machinery designed for right-handed people – with stop buttons placed incorrectly – are often overlooked.

In some sectors, such as construction or agriculture, you will need to reduce the residual, inherent hazards of the task as much as possible before relying too heavily on personal protective equipment (PPE) for individuals.

Security

Shared premises may present additional hazards associated with safety or security of common areas, such as lobbies, stairs, or escape routes. It is particularly important to consider processes of others in shared premises, proximity, and additional fire hazards introduced by their activities.

Consider the impact of the following legal requirements:

- Systems for monitoring location and state of materials; records of disposal
- Damage or theft of materials, goods during production or distribution
- Safety of goods in transit; storage facilities
- Unauthorized access to commercially sensitive or extremely hazardous areas
- Provision of appropriate training and support for staff to deal with potentially violent situations.

Consider areas where staff may be particularly vulnerable at certain times, such as opening in the early morning or locking up at night, when working alone, or when moving money and valuables around the site.

Competition

As information technology (IT) and Artificial Intelligence (AI) systems rapidly become more complex and sophisticated, so too do the means to illegally access other systems. This therefore represents a risk factor, whether it relates to medical or sensitive personal information about clients/patients/workers, or financial data of individuals.

Competitive issues to consider include the following:

- Minimum/optimum/maximum production schedules and ordering systems
- Industry image positive or negative; impact of bad publicity attached to other firms in locality or industry
- Uneven distribution of legal compliance amongst main industry players; inappropriate use of schemes to restrict access to suppliers

- Poor market reactions and subsequent loss of confidence by stakeholders
- Negative public perception of company as an employer if there is poor safety or health performance.

Apart from obvious short-term fashion trends, there are longer term life-cycle risk factors for most products. For example, the licensed trade and restaurant industries have been particularly badly hit in recent years, but coffee bars have seen a massive rise in popularity in the last decade.

Finance

Government levies introduced have an impact on high energy-users, and the "polluter pays" is the strongest message coming through in regulations for environmental protection.

Processes and substances in use for many years have later been found to be hazardous to workers and others, such as wood dust, and subsequently led to litigation. While no one can foresee whether this will be an issue, there may be some indications evident now. For instance, vapours given off during certain stages of flexible plastic processing may be considered just an irritant now but may later be classified as harmful.

Consider financial implications in the following areas:

- Research and development costs to use substitute materials effectively; staff retraining
- Cost of wages and other payments; ratio of wage bill to sales; profit-sharing schemes
- Cost of replacing outmoded or inefficient plant and equipment
- Insurance costs and potential for payouts in future if current acceptable processes are found to be hazardous.

While there has been a significant shift towards automated sales transactions, which are now seen as the norm, there are still a few cash-based businesses within leisure or retail. Risks associated with the use of cash can, therefore, be considerable for some firms, including potential for injury to staff as well as loss of money (often uninsurable). International pressure to shift to automated money transfer is not always viewed positively in all sectors and regions.

2.3.4 PLANNING AND POLICY

Planning

This includes planning at the strategic management level as well as the practical operations level, with all the other elements feeding information back to this stage and priorities for action being decided. For example, much of the work over recent years to develop tools to help businesses manage health and safety risks is due to this element being given a much lower priority than financial issues. Sometimes this may be justified, but recent shifts in the emphasis of regulations covering the workplace mean that all risks must be considered equally. Irrespective of the legal requirement, the aim here is to make sure you have a full picture of what is happening in your organization.

Questions at this stage include what is the purpose of the planning activity, who is involved in the planning process, either internal or external to the firm, and how will priorities for action be decided?

Employment

Larger organizations may be able to accommodate the absence of key staff members for some time, although the costs might be substantial. This is more difficult in smaller organizations, those that are traditionally female-dominated (with potential maternity absence), or in larger concerns with high levels of absence in certain divisions.

As firms expand, direct contact with each operational element of the business reduces, and delegation of authority and responsibility is necessary. Mergers and takeovers may also bring together business units with different beliefs and cultures, making it difficult to take the necessary holistic view of employment issues.

Factors to consider include:

- Insufficient input from all levels within the firm; exclusion of some groups of workers, especially those working on shifts/off site/travelling around from site to site
- Little understanding of the planning process and little commitment to the findings
- Balance between needs of different sites or business units
- Lack of knowledge or understanding of issues at operational level by senior management
- Lack of knowledge and awareness of current legislative requirements
- Ability to predict future skills needs of firm.

Legislation

Regulations are becoming more complex and far-reaching, with more scope for conflicting requirements from different enforcement bodies on the same issues, such as health and safety, fire, and planning requirements in food preparation sectors.

Consider the impact of the following legal requirements:

- Deciding priorities for action that acknowledge the legal requirements on business; time scales involved; amount of record-keeping and monitoring activities required by law
- Changes to licensing requirements for processes/ procedures/ qualifications
- Different legal requirements at regional local authority level and between countries
- Conflict between different enforcement bodies on acceptable actions required to comply with the law
- Changes to law that require alteration to premises; problems associated with age/type of premises and ownership status.

Security

- Access to relevant data to assist with planning
- Inappropriate use of data, either internally or externally.

Competition

The "boom and bust" cycles experienced internationally leave some negative equity situations, substantial losses on share values in some sectors, and potential problems for those responsible for making financial borrowing or investment decisions.

Regular downturns in economies worldwide have led to considerable financial damage for many firms. For instance, as the cost of fuel escalates, shortages in supply occur, and transport of goods is disrupted. This affects many industry sectors, and firms with little or no contingency planning in place are at a distinct disadvantage.

Global warming, regional and national transport policies, and increased flooding have all had a negative impact on infrastructure, representing a risk factor for distribution and delivery of goods.

Rapid advances in information/communications technology mean that it is difficult to realistically plan for future requirements.

Cybercrime is a major issue for managing both security and competitive risks.

Competitive issues to consider include the following:

- Correctly identify consumer trends and product life-cycle stage
- Awareness of future investment plans of competitors
- Developments in the industry; technological developments; access to relevant research
- Ability to access relevant skills and expertise to assist in planning process
- Relevance and sufficiency of available data; timeliness; volume of data to be collated and fed into planning process.

Finance

- Resource implications to support proposed plans for future
- Cash flow and access to necessary funds; opportunity cost of investments; prioritizing.

Many firms go bankrupt with healthy order books for the future but a critical shortage of funds in the short term – the cash flow is crucial. Late payment by businesses has been blamed in the past, and various national governments have introduced measures to try and alleviate this to some extent. However, the payment systems of the firm and its clients/suppliers are still a financial risk factor of all planning activities to consider.

- Rapid depreciation of some assets
- Government actions in future that have impact on long-term planning programme
- Base rate/interest rates for borrowing/tax rates and incentives
- Exchange rates/strengths and weaknesses of world currencies
- Consumer spending trends at home and abroad; inflation; regional upturns or collapses
- Uncontrollable events external to firm – for example, fuel price rises or shortages.

Policy

This may have been placed first on your own list on the basis that theoretically this should be the starting point. However, in practice this is not necessarily where people begin, certainly not in smaller firms where practical consider- ations dictate many of the subsequent policy decisions. It is, of course, a critical element in developing strategies that will enable the policy aims to be met.

A more realistic scenario is one where various elements of our risk factor lists feed into policy discussions and decisions, thereby making it a more dynamic element. Clearly such policies must be developed so that they can co- exist easily with others, and to have the greatest impact they must build on feed- back from the other elements identified. In addition, the question will always be "how do these policies affect the identified risks?"

The following lists do not include details of how you can word the statements as this will come later. For now, the list highlights the main hazards you have identified in the previous part of the chapter using the same headings as before.

Employment

- Involving and consulting with employees/workers on issues that affect their work
- Providing sufficient resources, information, and training to enable them to do their job effectively
- Ensuring equal opportunities for employment and advancement in the company, irrespective of gender, ethnic origin, religious beliefs, age, and disability
- Safeguarding workers from bullying or harassment in the workplace
- Setting targets for individual, team, and division achievement that are achievable, measurable, and realistic
- Providing adequate supervision and management at all levels of firm, with regular feedback on work
- Providing relevant discipline and grievance procedures.

Legislation

- Produce a product or service that does not jeopardize the safety and health of others or the environment
- Ensure plant and machinery are maintained properly, good housekeeping standards are set and maintained, and adequate facilities are available for those on site
- Provide a safe and healthy work environment for people; provide adequate protection against the risks of injury, harm, or damage resulting from work activities or fire; ensure the protection of vulnerable groups of workers, including lone workers
- Provide a "clean air" environment; no alcohol or illegal drugs are allowed on working premises; policy on drink/drugs and drivers
- In the event of a fire, safety of people will take precedence over damage to property.

Security

- Individual right to security of personal records and restricted access to confidential information; permission from individual to pass information to third party
- No commercially sensitive information to be passed to others outside the company
- Protection of designs
- Identifying and reducing opportunities for workers and others to be placed in personal danger of attack (such as locking-up/paying cash to bank)
- Policy on theft from company, clients, or other workers.

Competition

- Purchasing less hazardous and more environmentally friendly materials where possible

- Using local suppliers within x miles radius of business unit
- Appropriate use of third-party certification schemes to ensure quality of supplies, without introducing unfair and restrictive requirements
- Ensure any industry standards and codes of practice are adhered to
- Ensure appropriate record-keeping systems are in place and maintained correctly
- Establish and maintain effective financial controls to safeguard stakeholder interests, based on relevant legislative requirements.

Finance

- Ensure proper financial systems are in place to meet the firm's financial commitments to workers, suppliers, shareholders, and others.
- Provide sufficient resources to ensure that all legal requirements are met satisfactorily.
- Policy on late payment of bills; ensure adequate internal controls to meet all obligations.
- Establish systems to reduce potential for fraud within the organization.
- Provide sufficient insurance cover for insurable costs; consider how to cover any potential uninsurable costs.

2.4   SELF-REFLECTION QUESTIONS

- Are there still some areas to follow up, such as tasks or events that only happen once or twice a year?
- How have you made sure that all groups of people across the business have been able to help with this part of the risk assessment process? Are there any groups of workers that are still to have an input?
- Does your policy list cover all the hazard groups?
- What were the easiest hazards to identify and why was this? Are there specific hazards associated with your industry and, if so, what are they?
- How far do the hazards identified show where there are likely to be extra controls needed?

Keep these points ready to consider in the next two chapters to help identify potential harm and any new controls you need to put in place.

3 Evaluating the Hazards

3.1 EVALUATING THE HAZARDS IDENTIFIED

Clearly, the previous section is an extremely comprehensive list of the potential hazards or risk factors that any type or size of firm could find. While there are likely to be elements of each individual item that applies to any organization, the assumption is that at a business unit level, the scope of factors will be much smaller and therefore more manageable.

As we said earlier, external factors could have a higher or lower impact on the risks you face and so may need greater attention than mentioned here.

However, it must also be noted that this is only the starting point for evaluating the risks to the organization, and that it is important to spend time at this stage to ensure full coverage of the potential risk factors. It is then much easier to justify removing, or combining, some of these elements at future stages of the process to control and manage the risks effectively.

3.2 WHAT RESULTS ARE LIKELY FROM EXPOSURE TO THESE FACTORS?

So far we have identified "hazardous conditions, properties and processes that could potentially cause harm, injury or damage" as noted at the beginning of Chapter 2, and to some extent considered what form of harm, injury, or damage could result.

It is important to look at the range of hazards or risk factors in detail in order to get a clearer picture of what results of exposure to the hazard could be; how serious the resulting harm, injury, or damage might be; and who is most likely to be affected by it. In some cases, this will be specific to individuals at the shop floor or office level; in others it will directly impact the business unit or division; and in some other situations the resulting damage could affect the whole organization or geographic location where it operates.

Potential harm, injury, or damage results from:

DOI: 10.1201/9781003405641-5

- Employment factors – the inability to recruit suitably qualified and experienced staff, leaving the company vulnerable to poor workmanship and quality problems. In some industries, an ageing population means the workforce is made up of many nearing retirement age, but with few young workers entering the industry. Working conditions can lead to low morale or commitment of workers, high staff turnover, and potential for needs of vulnerable groups to be ignored.

- Legislative factors – increased costs/time involved in compliance with employment legislation, especially if internal knowledge of the law is limited. Personal injury to workers/customers from unsafe actions, harm to individuals or local environment from incorrect handling of products, leads to increased calls on the cost of insurance cover. Changed priorities of local authorities in enforcement actions or planning decisions can involve increased investment by company. Insufficient safeguards against the start or spread of a fire can cause potential damage or injury over a considerable part of the local area.

- Security factors – growing in importance for many firms as the potential for harm can be considerable, ranging from leaking sensitive information that can damage future investment plans to loss of personal data of workers or customers. Data protection is crucial given the rapid growth in cyber-crime and the greater reliance on IT systems by customers, suppliers, and the business itself. As theft, burglary, and violence to staff increase in some regions, so does cost of physical guarding and alarm systems, as well as potential claims for injury or loss.

- Competitive factors – declining industry sectors, the need to compete more globally than previously, and investment costs of research and development all represent significant risks. An additional competitive factor is loss of sales, profits, and market share as customer expectations change, particularly the major shift to online ordering. On the other hand, inadequate premises that are poorly located means customers are discouraged from visiting the site. Consumer pressure continues to grow for more environmentally friendly products, recognition of need for environmental protection amid global warming concerns, and demand for processes/services that need significant changes to how you operate. We have seen major global upheaval and conflict outside your direct control forcing a rapid change in processes and procedures – for example, the unprecedented global pandemic of 2019–2022, and restrictions in transport or supply of materials from the Eastern bloc.

• Financial factors – increased cost of operating the business, recruiting and retaining relevant skilled staff, has to be balanced by other costs. Late payment is critical to cash flow and can significantly increase costs of borrowing in the short term, as reliance on limited customer base leaves the firm vulnerable to the loss of just one or two major customers. Overall increased emphasis on protection and compliance with a wide range of legislative requirements takes funds away from the direct purpose of the organization, that is, to provide a specific product or service for the customer. External perceptions about the firm can have major financial impact when stock market values are volatile.

Keeping these five risk factors in mind and considered against each of the 10 Principles, the process will inevitably have led to repetition or overlap of factors although we have tried to keep this to a minimum. Having carried out this vital stage at such depth, the risk factors can now be grouped together before considering the potential impact of the hazards. Tables 3.1 to 3.4 summarize the risk factors identified earlier, breaking them down into more manageable chunks. This is just one option for summarizing the findings, but it does help to see where hazards and risks overlap.

The breakdown of headings in Column 1 reflects the type of hazards found in the individual firm, so this is just an example of how it might be done. However you choose to present your findings, it is useful at this stage to keep to the four headings related to physical properties, people elements, processes, and management issues.

Note that a final column is useful for pulling out factors that occur across the group, such as skill base in the local area, or facilities available, and may help to emphasize significant factors that appear regularly and need closer attention. In the example shown here, the factors relate to a typical manufacturing plant.

3.3   WHO IS LIKELY TO BE AFFECTED?

Having completed this stage, a further column, Column 5/6, would help to identify which individuals or groups of people are most likely to be affected by the potential risks specified. Remember to include reference to:

• Individual workers at all levels within the organization, including those on part-time or temporary contracts; shift workers; home or mobile workers; people not based permanently at just one site; drivers and transport workers; agency staff
• Other workers on site not necessarily employed by the firm directly or other commercial occupiers of premises; visitors on site, whether authorized or not
• Customers or clients, both on and off site; suppliers
• Shareholders and other stakeholders in the organization; directors and senior management staff
• Investors; finance providers such as banks; insurance providers
• Residents, industrial and commercial users in the vicinity; local wildlife habitats and the environment.

TABLE 3.1
Risk Factors Identified – Group A: Physical Properties – Premises/Product/Purchasing

Risk Factors	Premises	Product	Purchasing	Common Features
Employment location/ skills/ demographics	Access	Age of workforce (10% to retire in 10 years)	Distribution routes	Location and distribution channels
	Size/capacity	Mid-life-cycle stage of product	Few local suppliers	
	Facilities	Distribution routes	Poor communication link with production	
	Car parking Image	Local skill base		
Legislation: employment/ health & safety/ environment/other	Lack adequate facilities	Fire risk re-use of materials	Need to source safer materials	Inadequate storage
	Age premises/stairs/ corridors	Need for safer substitute materials	Poor facilities for drivers	facilities
	Unsafe structure	Product safety features	Insufficient storage	Need for new sources safer supplies
	Heating/lighting Noise: close to residential area Restrictive planning requirements	Disposal when obsolete Amount/type of packaging used	Disposal of waste	
Security	Safety in local area	Copyright protection	Perimeter fence damaged and stores not fully secured	Local security measures
	Safe storage of vehicles	Storage and warehousing	Checking out system not working	Inadequate
	Warehouse facilities poor	Access to data pilfering small parts		Internal security systems

Competition: Industry/consumer/internal	Unreliable power supply locally Poor local amenities Growing sector/no purpose-built premises Image: nearby derelict shops	Decline in traditional Industries locally Short-term life span so constant up-date needed New R&D facilities needed	Turnaround time for deliveries Increasing Packaging Limited access to eco-friendly substitute materials	Packaging and disposal Image/R&D
Finance: internal/external	Increasing maintenance costs Cost relocation of key staff Cost insurance and security Incentives for staff	Increasing cost materials Cost of scrap High cost packaging price Increased transport costs	Increased cost of transport Late payment by customers Our late payment restricting supply from main supplier	Cost of materials Cost of insurance Fuel costs

TABLE 3.2
Risk Factors Identified – Group B: People Elements – People/Procedures/Protection

Risk Factors	People	Procedures	Protection	Common Features
Employment location/ skills/ demographics	Low levels of skills locally	Worker inexperience	Flexible working and need for staff cover for absence	Staff turnover and absence
	Need for training	Recording procedures not fully working	Unauthorized access to site	
	High-level temporary work contracts used			
Legislation: employment/ health & safety/ environment/other	Welfare facilities limited	Need to substitute safer	Worker protection measures	Noise levels
	Health concerns regarding light/noise levels	materials	Need to update H&S training	Hazardous waste and scrap
	High staff turnover	Use of PPE	No proper consultation system in place	Training and supervision
		Health surveillance in paint shop	Protection of young workers	
		Maintenance	Noise levels	
		Disposal of hazardous waste	Fire hazards	
Security	IT systems not secure	Data protection	Potential arson at edge of site	Vehicles
	Authorized access systems not fully working	Need to monitor Internet use	Unauthorized use of vehicles	IT systems
		Virus control		
		Secure storage of vehicles		

Competition: industry/consumer/internal	Competition for skilled staff	Compliance with quality system in workshop	Image of premises	Competing for staff
	Retaining trained staff	Need to replace two machines Speed of finishing goods	Protecting access to supplier base	Quality of product
	Few young people in industry	Stock control	Competing for skilled staff	
Finance: internal/external	Cost of training	Cost of waste/scrap	Cost of replacing old equipment	Employment costs
	Rising cost of employment	Cost of rework for low quality	Need noise reduction measures	Investment in plant and machinery
	High sickness absence over 12 months period	Increase in customer returns	Cost supply of PPE for new staff Increasing wage bill	
			Access to pensions 2001 onwards	

TABLE 3.3
Risk Factors Identified: Group C – Processes/Performance

Risk Factors	Processes	Performance	Common Features
Employment: location/ skills/demographics	Lack skills in new workers Little choice about process Some work repetitive	Targets not set consistently across firm Data collected incompatible across firm Discipline procedures too heavy-handed	Way work is measured against targets Skills
Legislation: employment/ health & safety/ environment/other	Need to replace some equipment with safer versions Potential health problems in paint shop Manual handling Safe disposal of waste materials Fire hazard Noise levels	Internal procedures for dismissal not used correctly Reports of diseases or injuries not collated No one monitoring sickness absence Need to review power use regarding any climate levy	Replacing equipment Use of existing systems
Security	High damage levels to portable equipment Use of trucks on-site Power supply and computer-aided processes	Close-knit local community and access to sensitive information	
Competition: industry/ consumer/internal	Becoming out-of-date Need to use new materials Transport costs	Inconsistent against MSS criteria Poor industry benchmark rating	Rating against internal and external criteria
Finance: internal/external	Insurance costs Costs of rework Cost of sales Stock control and cash flow	Cost of absence Potential cost of levy Potential loss market share	Cost of poor quality

TABLE 3.4
Risk Factors Identified: Group D – Planning/Policy

Risk Factors	Planning	Policy	Common Features
Employment: location/skills/demographics	Access to temporary staff involving mobile staff Little current knowledge of future skills needs Changes needed to employment practice	Existing employment policies not adhered to No policy on disability or rehabilitation Discipline and grievance procedure too cumbersome	Employment practices identify future needs
Legislation: employment/health & safety/environment/other	Use of new materials in planned production Time available to existing staff to find new suppliers Lack internal competence on H&S	Need for smoking policy Gaps in compliance with H&S regulation Policy on disposal of obsolete goods	Urgent need to ensure compliance with legislation
Security	Poor base of relevant data to draw on	Security of personnel records	Access to data
Competition: industry/consumer/internal	Access to relevant market data, internal and external Technology expertise	Need for eco-friendly purchasing Market awareness	Market analysis
Finance: internal/external	Minimal value remaining in existing plant Substantial investment required, both short- and long-term	Confirm policy on late payment Returns needed on capital invested	Investment

This may be clearly restricted to a specified location, business unit or division.

> For Example: Storage facilities at a plant are clearly inadequate for the growth in production activities taking place there, leading to increased exposure to hazards associated with unsafe racking or movement of goods. There are increased waiting times for unloading deliveries, wasted time due to ineffective recording/stock control systems, and a drop in quality compliance levels.

In other situations, particularly health and safety or employment issues, there will be individuals who are most at risk.

> For Example: The sales force relies on vehicles being safe and roadworthy, fuel readily available, and roads being passable. Fuel shortages, major road works, and flooding increase the stress levels amongst these workers, and greater incidence of "road rage", plus longer working hours due to hold-ups.

Purchasing or performance factors are likely to impact throughout the organization, especially where several business units are widely spread geographically. Changes to working practices, such as working from home, hybrid or flexible working for all or part of the week, must also be included in deciding hazards or potential harm, though not always as easy to identify.

There are two more elements that need to be considered to fully evaluate the potential risks to the business, that is, the level or severity of harm that is likely to occur, and the likelihood that it will happen. There are many ways to approach this stage, and the next chapter looks at some of them.

3.4 SELF-REFLECTION QUESTIONS

- What are the major employment factors you have identified?
- What are the biggest risks of harm from legislative factors?
- How important is potential harm from security factors?
- Which competitive factors have the most potential to cause harm to your business?
- Are financial factors more or less than you expected to find?
- Have you identified all the people most likely to be affected by these factors?

4 Evaluating the Risks

4.1 RATING THE EXTENT OF POTENTIAL HARM/DAMAGE

It is impossible to suggest that there is just one correct method for rating potential harm. While we have produced some very detailed notes in the previous chapters, it is still primarily a subjective exercise to allocate some form of rating to all the different elements that constitute a "risk" to the business. In addition, anyone who wants to see that such an evaluation has been carried out will come with their own set of objective and subjective measures against which to validate your version. Having said that, it is valid to produce an evaluation of risks based on the following:

- A comprehensive base of knowledge about the potential hazards or risk factors
- Knowledge of the context in which the business operates (external factors often outside the direct control of the company)
- A rating system that is as simple or complex as it needs to be.

You can choose to allocate a numerical value to these evaluations, or to make a judgement based on criteria such as HIGH/ MEDIUM/ LOW, whichever you find is the easiest.

In a health and safety context, the possible severity of harm associated with a hazard is generally quite straightforward. There is a lot of information out there (wherever you are based) to illustrate the type and severity of harm likely to occur from exposure to specified hazards.

For example, physical injury to upper limbs is associated with the use of machines used in wood-working shops, plus the development of nose/throat/lung cancer from exposure to carcinogenic hard-wood dust particles. In this situation, identification of individuals most likely to be exposed to such hazards is reasonably obvious, as is the location.

How harmful exposure is to health-related hazards is sometimes more difficult to establish, particularly those with long latency periods between exposure and appearance of symptoms. For example, exposure to some of the chemicals used in metal manufacturing processes during the 1950s and 1960s only appeared as long-term damage after many years. Previous use of substances or materials is an additional

DOI: 10.1201/9781003405641-6

factor that needs to be considered against "Product" or "Process" in earlier chapters, especially where they have subsequently been reassessed as hazardous.

In the context of health and safety, typical ratings for potential harm include:

i)   "Slightly harmful" or "Low" rating – such as minor or superficial injuries that might or might not require first aid treatment; levels of noise or other emissions at current minimum levels allowed by law.

ii) "Harmful" or "Medium" rating – such as serious sprains or fractures, burns, or concussion that result in lost time or require a hospital visit, potential for harm to some vulnerable groups of workers.

iii) "Extremely harmful" or "High" rating – including potential for major injuries, fractures, irreversible chemical damage; significant hearing damage; and, of course, death.

Fire or environmental risk factors need to be considered differently given the type of harm or damage that can occur. How severe the likely harm or damage is not so easily related to individuals carrying out specific tasks as it is often more invasive or all-inclusive. There are similarities between the way a fire might spread and a chemical spillage, for instance, where the impact can be felt over a wide area. Typical ratings in this case could be:

i)   Small scale, slow-release localized – can probably be tackled safely in the early stages by a competent person

ii) Small scale, localized to start with but potential for rapid spread (either fire or spillage)

iii) Likely rapid spread and/or potential for producing toxic fumes or smoke affecting a wide area

iv) Instant spread possible over a wide area, perhaps via a spread of chemicals, combustible materials, or dusts

v)   Significant potential for explosion.

In these situations, the impact may well be on others outside the organization, rapid and localized, slow, and widespread, or indeed spectacularly widespread in the event of an explosion. If preferred, the ratings could be combined as:

•   (i) and (ii) equivalent to the "Low" rating
•   (iii) equivalent to "Medium" rating
•   (iv) and (v) equivalent to "High" rating.

Severity of harm likely to arise from security hazards depends largely on the size and range of business activities undertaken, as well as the industry sector. The level of sensitivity of data held will determine the extent of harm likely if it is lost. Cybercrime is a major global concern whether you are a tiny micro firm or a large multi-national. Rapidly becoming more sophisticated, the impact can be devastating in both the long and short terms. Not easily foreseen or controlled, it certainly constitutes a high risk to any business.

The loss of a few hundred pounds or dollars from the till of a local shop will have a much greater impact on the business than the same amount lost through staff pilfering in a larger organization. However, such losses can have a wide-reaching impact on other risks facing the firm, certainly financial and competitive risks. Ratings in this context might include:

i)   "Low" rating – inconvenient loss, little impact on other elements of the business, already costed in

ii) "Medium" rating – not immediately or easily recoverable, not fully covered by insurance; impact on other elements of the business; some public embarrassment

iii) "High" rating – severe impact on the financial viability of business; far-reaching impact of bad publicity; non-recoverable loss.

Competitive or financial risk factors are often less easy to quantify for potential severity of harm likely to occur. This may be due to the potential for misinformation or incomplete data on which to base analysis, the need for optimism (whether well founded or not!), and the impact of events outside the direct control of the organization. Of course, management theory has developed to take these points into account, and various financial modelling and theoretical tools are available to assist in this, such as the RMF in the US.

While systems will already be in place to evaluate these risks in big business, possibly more so than for operational risks, a micro or small firm is unlikely to have access to such facilities. A similar approach to previous risk ratings is included here as a starting point.

Global trading and widespread commercial use of the internet may not necessarily represent a significant problem, as you may be able to tap into a much wider customer base than what was previously the case. With fewer industries relying on a local customer base in regular face-to-face service provision, costs of distribution of materials, goods, and practical logistics of fulfilling orders now represent a different range of risks.

If larger clients insist on all suppliers obtaining third-party certification to demonstrate compliance with specified standards, this might attract a "Medium" or "High" rating against severity if the cost of such certification becomes prohibitive. This is particularly relevant when new trade deals are negotiated. Ratings might include:

i)   "Low" rating – slow change of fashion trends for product; established firm with few major competitors locally; relevant standards already part of normal operations

ii) "Medium" rating – increased compliance requirements from enforcement bodies leading to additional investment; external factors such as fuel shortages or high fuel price increases leading to significant disruption to delivery mechanisms, staff travel – more severe impact in some rural locations

iii) "High" rating – traditional industry in decline; major employers leaving region leaving and thereby leaving local support industries in critical position; lobby groups severely restricting ability to operate or attract necessary funds (such as environmental protection groups).

More so than in other areas, financial harm will depend entirely on the way the business is structured, funded, and organized. In a guide like this, aimed at the smaller businesses, we cannot cover every type of structure. However, the severity of potential harm will still relate to:

- Cost of operations relative to sales
- Ratio of borrowing to value within the business – an issue no matter what size the firm is
- Cost of borrowing, access to funding when needed, interest and exchange rates outside the firm's control
- The increasing cost of insurance cover coupled with the reduction in the level of cover provided.

These elements are compounded in some instances by the expectations of shareholders, perceptions of financial markets, and potential loss of confidence by other stakeholders. The relevant ratings of "Low", "Medium", and "High" can be used to identify how severe the potential impact will be on the organization, from having to economize at the lower end to consideration of divesting part of the organization or selling up altogether, or indeed insolvency. As with other factors, such as legislative issues, the potential for fines or penalties being levied may be a factor to include in the evaluations. These will all relate directly to where you operate the business from.

The immediacy of impact on the business will need to be an integral element of the severity of harm assessed under each of the factor headings. With some risks, the fact that it might be two or three years before the impact is felt adds a further dimension to the evaluation, even more so if the time scale is further into the future.

A separate, but just as significant, issue to include in the calculation is whether the potential impact will be represented by a short-term "blip" in operations, has ongoing long-term effects on the way the firm performs, or requires a fundamental change in structure or approach. Many issues seen as specific to a few individuals can have far-reaching effects on other workers or outside observers.

Although more thought is required to allocate a number score rather than a low-to-high rating, it does provide the opportunity to consider sliding scales for severity of potential impact, immediacy, short-/long-term effects, and a spreadsheet basis for comparing the findings later. A useful scoring system is one based on either a maximum 30 score or perhaps 50 score, allocated on the following basis:

i) Low rating – up to 10 score
ii) Medium rating – between 11 and 20 (or 11–40)
iii) High rating – between 21 and 30 (or 41–50)

4.2 EVALUATING THE LIKELIHOOD THAT HARM/DAMAGE WILL OCCUR

We now have a very comprehensive picture of potential hazards to the business, and undoubtedly an extremely depressing picture of everything that could possibly go wrong! Fear not! The next stage is to assess the likelihood that this harm, injury, or damage will occur. Clearly, there are already many forms of control in place to ensure the potential damage is not so severe, whether these are physical controls, safe working practices properly supervised and monitored, contingency plans in the event of disruption to usual transport routes, or other valid means to reduce potential harm. Some of these controls will be considered in the next chapter, but for now the "likelihood" factor will need to be assessed against some relevant criteria.

For instance, the more people exposed to a specific hazard and the longer the exposure time, the greater the likelihood that harm or injury will occur. This might be due to:

- The compounding effects of exposure over time, such as harmful substances acting as sensitizers
- Complacency and carelessness as activities become a familiar habit
- Increased exposure to high-risk activities such as driving where the individual is not necessarily in sole control of events.

For Example: A 45-year-old owner of a sawmill had part of his hand severed by a bandsaw, despite working in the plant for 25 years. He noted "It's easy to become a bit blasé about working practices after so long in the job".

On the other hand, there may be a greater likelihood of harm due to inexperience of the individual, lack of knowledge and awareness, or forgetfulness if the activity only occurs infrequently. Gradual breakdown or wear-and-tear of equipment/machinery/structural features/infrastructure will increase the likelihood that harm or damage will occur, as will periods of close-down for repair and maintenance.

However subjective the evaluation might be, you should now have a reasonable picture of:

- Hazardous conditions/properties/processes that could potentially cause harm, injury, or damage
- What this harm, injury, or damage might be; who could be affected; and how serious the result of exposure might be
- The likelihood that such a harm, injury, or damage will occur, taking into account any control measures that exist.

Chapter 5 includes examples of different ways to assess the potential risks using three or five categories, or a numerical score.

4.3   SELF-REFLECTION QUESTIONS

- Have you used the third or fifth rating scale for considering potential harm from hazards?
- Did you find it easier or more difficult to decide ratings for competitive or financial risks?
- Once you decided the ratings, did they help when evaluating what the likelihood of harm would be?

Part 3

Controlling the Risks

5 Risk Controls

5.1 POSSIBLE CONTROL MEASURES

Various ways of controlling risks have been mentioned already, and clearly there are as many ways to control or reduce the impact of a hazard as there are types of hazard. There are also different interpretations of the term "control". In this instance, we refer to a control measure as

> An action/ device/ strategy intended to eliminate/ alleviate/ reduce the negative impact on the business or individual of a situation or event.

You are already likely to be familiar with most controls commonly used, although it might only be a direct experience in the context of health and safety controls, or of financial controls. It is therefore worth reviewing the different categories of controls that could be used in an individual firm.

The following headings are used for convenience, some things falling clearly into one category or another, others included in a category on a seemingly arbitrary basis.

1. Physical controls

    *   Health & Safety – wide range of physical controls such as guards, rails, and barriers; lifting and handling equipment; fail safe systems and stop buttons; personal protective equipment (PPE) such as hats, goggles, boots, gloves, masks, harnesses, breathing apparatus; ventilation and exhaust systems; vehicle alarms and specified routes on site
    *   Fire – fire, heat, or smoke alarms; notices and warning signs; fire-fighting equipment; circuit breakers; restricted access areas or containers for flammable materials; fire doors; sprinkler systems
    *   Washing and decontamination facilities; tachometers; clocking in records
    *   Use of CCTV and security badges; time-activated locks; access barriers
    *   Security systems for access and use of IT equipment and protection of data

2. Behavioural controls

    - Relevant training provided at all levels of the organization; specific qualifications required to operate in some areas
    - Communication systems that reach all sectors within the organization; regular meetings to provide updates on progress
    - Individual responsibilities and boundaries of authority clearly identified and agreed provision of adequate supervision
    - Incentives, reward schemes, and other methods used to motivate and encourage staff, including internal promotion and advancement possibilities
    - Culture, including attitudes towards taking rest breaks and holidays when due

3. Organizational or procedural controls

    - Planned maintenance programmes; regular checks on equipment and machinery as the norm; safe working procedures and "Permit to Work" systems in place
    - Regular health surveillance where necessary; hearing and sight tests provided; review of work patterns to reduce repetitive movements
    - Emergency procedures in place and tested regularly; contingency plans (for spillages, for example); washing and decontamination procedures adequate for the type of work
    - Consultation with employees via their union or worker representatives
    - Use of security passes and authorization arrangements complied with; culture of enforcing internal rules (such as wearing hard hats/hearing defenders) as the norm
    - Appropriate recruitment policies and practices to ensure compliance with equality of opportunity legislation; staff screening systems; employment contract agreements
    - Adequate allocation of resources
    - Signatory procedures in place for transactions; log on and password procedures; systems to safeguard lone workers
    - Use of Management System Standards or industry standards; use of Approved Codes of Practice (ACOP)
    - Collection of data at regular intervals; senior management/board-level commitment to protecting people from risks
    - Relevant insurance cover.

The list is not exhaustive but highlights how difficult it is to categorize elements that overlap and interact with each other so closely. There may be other actions, devices, or strategies that act to reduce the potential impact of risks identified in any one organization, or a combination of measures that are necessarily much narrower than those listed here. In any event, control measures that exist can be considered alongside the risk factors identified earlier, either within the framework of the five risk factors:

- Employment
- Legislation

- Security
- Competition
- Financial

or against each of the ten headings used for the 10 Principles:

1. **Premises**
2. **Product**
3. **Purchasing**
4. **People**
5. **Performance**
6. **Protection**
7. **Process**
8. **Procedures**
9. **Planning**
10. **Policy**

Having considered this substantial amount of data collected, there is room for a critical look at how valid these findings are to make it easier to "make judgements about adequacy of controls in place and identify gaps in provision" (Chapter 1), before deciding priorities for future actions that might be needed to correct the situation.

5.2 SYSTEMS OF CONTROL POSSIBLE

As a business grows, it is inevitable that more formal systems than generally associated with SMEs will be established. Experience suggests that around 50 employees is the point at which the entrepreneurial, flexible management approach no longer works effectively (on the assumption that it ever did!). When carrying out such a far-reaching analysis as this one, it is important to confirm that the data is relevant, accurate, and up-to-date. Even more important is the need to be critical about whether the existing controls are in place or are assumed to be working because the rule book says they should be.

Therefore, the question posed is as follows:

How effective are these controls, and what systems are in place to ensure they are appropriate, working, adhered to by everyone, still effective?

Use the following summary as a prompt to consider this question more closely.

a) Employment controls:
   - Recruitment: are job and personnel specifications relevant to the jobs as they currently are, rather than how they used to be? Are specifications appropriate for bringing in the correct skills and expertise required now and in the future? Are future skills identified in plenty of time to develop them in-house or recruit from outside? Who makes these decisions and on what basis? What flexibility of working patterns is available to workers?
   - Equal Opportunities: does the mix of workers reflect the make-up of the local community? Is there a mix of ethnic backgrounds amongst the

workforce/a mix of age groups/a fair balance of male–female workers in all job areas? Have efforts been made to accommodate applicants or workers with any special needs or support?

- Training: is training provided to workers across the organization? Are there restrictions on access to some training based on age or sex of the worker? Is this based on valid assumptions? Is there a full programme of induction training, including when people move from one site or division to another within the same company? What is the balance between internal/external/general/industry-specific training? What support do you provide for workers who wish to undertake their own training or studies?

- Communication: are adequate systems in place to ensure that proper channels of communication are maintained throughout the organization? Is there real commitment to proper consultation with workers? Are all groups of workers involved, including those on shifts or regularly working off-site?

- Supervision: is direct supervision appropriate, effective, and maintained? Have vulnerable groups of workers been identified for additional support if necessary? Are supervisors and managers trained adequately in this role, rather than just their area of technical expertise?

- Premises: how often are facilities reviewed to confirm that they are still adequate? What is the refurbishment programme timescale? Who decides on structural alterations or decorating designs?

b) Legislative controls:
- Compliance: how does the organization stay up-to-date with legislative changes? How are people notified of changes and how will they be affected? Is specialist advice available internally/externally/cross division? How is this monitored? Are legislative differences world-wide recognized fully in cross-border agreements?

- Health & Safety: how are results of risk assessments recorded? Is this procedure standardized across all divisions? Have people received training in carrying out risk assessments? What are internal procedures for reviewing and updating risk assessments? Are these carried out when new plant or machinery is introduced? Have vulnerable groups of workers been identified?

- Employment law: is the internal discipline and grievance procedure effective? Has the list of "gross misconduct" (often referred to as dismissible offences) offences grown so long as to include every possible minor infringement, affecting morale and motivation? Are provisions under Maternity Leave, Working Time, and other major legislation (as required in the UK) applied equally throughout the firm? Is there a history of cases being brought against the firm in tribunals, civil or criminal courts?

- Environment: are waste management systems effective? Are emissions checked as often as necessary? Are maintenance programmes established and adhered to? Are all relevant licenses obtained and conditions complied with?

- Records: are all relevant records complete and up-to-date? Is responsibility for completing records correctly allocated? Are records accessible or restricted appropriately? Is the data collected relevant and in a usable format? Is information passed to the relevant authorities according to legislative requirements?

c) Security controls:

- Physical security measures: are insurance providers satisfied that adequate security measures are in place; what is the record of claims, break-ins, damage to property? Are security providers vetted sufficiently? Are systems working correctly?

- Access: is everyone provided with correct authorization? Are new staff screened before they arrive (e.g. through police or social services records) if relevant to their position? How are computerized records protected? When was this system of protection updated and by whom? What checks are in place to confirm the protection works effectively? How are secure data sources accessed when authorized personnel are absent? Are systems in place to combat viruses or deliberate sabotage of data or cyberattacks?

d) Competitive controls:

- Information: how are customer complaints or returns dealt with? Have complaints or reported faults risen or declined in recent years? What trends have emerged and what reasons have been identified for them? What communication channels exist across and between companies?

- Marketing: how is internal and external market intelligence data used and at what levels in the firm? How is it collated and disseminated? What measures are used to monitor the results of market strategies? How and when are these reviewed and amended if necessary? How have ratios of marketing costs to sales or profits altered over the last 5 years and why? How effectively is the internet used for commercial transactions, especially social media?

- Standards: how does the firm compare with other industry players in benchmarking exercises? Are relevant industry standards used effectively across part of or all sections of the organization? How are changes to these requirements implemented? Are the principles of continuous improvement adhered to across the whole organization? If not, where are the gaps and are they acceptable? Is the application of international standards' requirements fully complied with, maintained adequately, and supported by organizational commitment? Have these systems been fully integrated with each other and any others that are relevant? Have differences worldwide been recognized without putting some nationals at a disadvantage compared with others?

e) Financial controls:

- Procedures: do accounting procedures fully comply with legislative requirements? Is the data generated available to the right people at usable intervals for them to make financial decisions, weekly/monthly/quarterly rather than after the end of the year? Are financial records fully compatible across the organization, whether based inside or outside the home country? Are records fully up-to-date at any given time in the financial year?

- Cash flow: is cash flow given sufficient attention as well as sales/costs/end of year profits? Are internal and external late payment procedures in place and working adequately? Does the firm also ensure it pays bills on the same principles of late payment to avoid restricted or delayed supplies? Is factoring used for collecting bad debts? What is the debt recovery pattern over the last 2 years?
- Insurance to spread the risk or liability: when was the last time insurance cover was renegotiated rather than just renewed when due? Is insurance cover adequate for new or different risks that may have arisen, such as new plant or machinery to replace the old, or exports to different countries from those cited in original Proposal Forms to insurers? Has value of insurance cover kept pace with increases in value of property? Is compulsory Employers Liability cover held? What is claims record – any patterns emerging?
- Capital: how are changes to exchange rates/bank rates/inflation/ borrowing terms/national or regional taxes monitored? How are stock market movements monitored and evaluated? Are investment reviews and evaluations carried out internally/externally/a mixture of both? How often, and is this satisfactory to ensure the best for the firm? Are compliance costs and penalties identified and included when financial planning is undertaken? Is access to technical and financial expertise available?

By this stage, none of these points will necessarily be different from what has already gone before, but referring back to them in each context reinforces their importance and ensures they are not forgotten. Although adding still more dimensions to consider when assessing risks, this will help to clarify elements of the business that have been neglected or perhaps under-rated previously and emphasize the need for a holistic approach.

If the organization is complex and a numerical rating is applied at various stages of the process, a clearer grading system should emerge, so making the next stage of deciding priorities for action that bit easier.

5.3   DECIDING PRIORITIES FOR ACTION TO ELIMINATE OR REDUCE RISKS

Listing factors in descending order of scores makes it easier to start the prioritizing process, and to illustrate how wide-ranging and complex the risks to the business are. However, the danger is that just those at the top of the list are tackled, leaving the lower-scored issues to be dealt with on a more ad hoc basis.

Alongside the ratings given to risk factors there are still more questions that need to be asked. Having assessed the likely extent of harm or damage, and potential disruption to business activities, the organization's ability to recover from the impact will be crucial to the prioritizing process. A critical look at the robustness of internal systems, and evaluation of their strengths and weaknesses within all areas, is a vital element of the subsequent decision-making process.

If a wide range of risks need to be addressed, then some form of prioritizing must take place to ensure that action is taken, within a given timeframe, to eliminate the risks or reduce them to an "acceptable" level. Less demoralizing than identifying lots of areas that need to be addressed is for nothing to be done at all to correct them.

If the original risk factors identified are collated with recurring or overlapping themes combined, a list of "primary risk factors" will emerge. These can then be plotted on a grid such as the example shown in Table 5.1 and should result in some form of scattered response pattern.

In this example, three divisions are used on each axis, resulting in nine categories ranging from:

- A "trivial" or low risk score of (**1**) representing unlikely or low-harm events
- Score (**2**) where an event is likely to occur although the resultant harm may not be great
- Score (**2**) where an event is unlikely, but if it did occur, the harm could be significant
- Medium score of (**3**) where an event is very likely, but the resultant harm is "trivial"
- Medium score of (**3**) where an event is likely to occur and the resultant harm or damage could be significant
- Medium score of (**3**) where an event is very unlikely, but if it did occur, the result could be extremely harmful
- High score (**4**) where an event is likely to occur, with results extremely harmful
- High score (**4**) where an event is very likely to occur, with significant harm or damage

TABLE 5.1
Assessing the Risks – Three Categories

	Slightly Harmful or Low-Level Harm	Harmful	Extremely Harmful
Low likelihood/ highly unlikely	Trivial risk 1 *******	**** 2	*** 3
Medium likelihood/ likely	******* 2 **********	** 3	4
High likelihood/very likely	3 ***	*********** 4	Intolerable risk 5 ******

Note: * = risk factors identified. The number of * shows an example of where one might find lots of points of concern or just a few. That is, the scattered response referred to earlier.

- An "intolerable" or high-risk score of **(5)** where an event is very likely to occur, and would be extremely harmful if it did – representing the need for urgent action.

Immediate action is clearly required for those risk factors that appear in the last category – that is, with a score of **(5)** – in that the potential damage could be devastating to the individual or business and is, indeed, very likely to happen. These factors require action within the short term, to reduce the likelihood that they will occur, to increase protection, and to reduce the potential impact if it does.

Risks that fall within the boxes showing a score of **(4)** also represent significant risks that may require urgent and serious consideration. Ultimately, the aim is to generate a significant shift of risk factor scores from the bottom right-hand box as close as possible to the top left-hand box.

The second example in Table 5.2 shows a more extended table based on five-by-five categories, allowing for a finer delineation of elements that impact the final position in the chart. Again, the bottom right-hand corner represents the most significant risks to the business, this time labelled as Intolerable in three boxes rather than just one.

If a scoring system of 1–10 is used to evaluate the potential impact of each risk factor, as in Figure 5.1, the emerging pattern shows the highs and lows of the range to help decide on priorities. Decisions can then be made about the balance between levels of scores, urgency of action required, and order of priority for action. In this case, a mid-range "Priority" line may appear as a natural divide.

TABLE 5.2
Assessing the Risks – Five Categories

	Slow Spread/ Localized	Rapid Impact/ Localized	Harmful	Rapid Impact Wide Damage	Explosion/ Severe harm
Few people affected	1 ***	1 ******	2	2	3 **
Highly unlikely	1 ***	½	2 **	3 **	4
Likely	½	2 ***	3 ***	4 ****	4
Very likely	2	3	4 ***	4 ***	Intolerable 5 **
Many people affected	3	4 **	4	Intolerable 5	Intolerable 5 **

Note: * = risk factors identified. The number of * shows an example of where one might find lots of points of concern or just a few. That is, the scattered response referred to earlier..

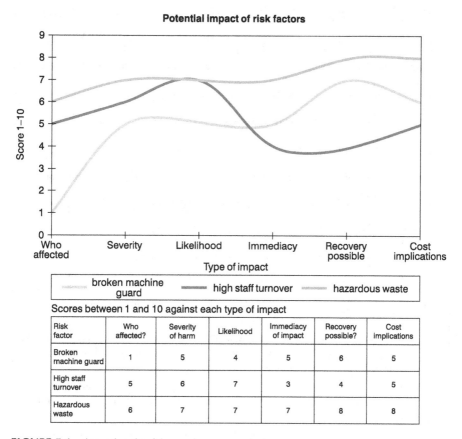

FIGURE 5.1   Assessing the risks – using a numerical score.

Finally, cost implications of not taking any action must be balanced against likely cost of dealing with the risk factors. In this case, it must be stressed that cost may indeed be a factor in the level of action taken to reduce risks but must not be the basis for a decision to take no action at all – certainly where legislative risks are identified. Over-reliance on comparing reasons for deciding priorities could also conceal the fact that a low total score incorporates one element with a very high score for severity to an individual. Action must, therefore, be taken to protect the individual from harm.

5.3.1   How Tolerable Are the Risks?

Assuming that the levels of risk identified are a fair reflection of what is happening in the business, you will have to decide which risks appear as "tolerable" with no further action needed at this stage. Note that a very low rating score does not mean the risk factor is unimportant. After all, the scoring system has been applied to risk factors evident in the firm, not just a hypothetical list of what might be present. Indeed, a large base of "trivial" risks can increase the rating for other factors such as motivation of staff, or quality of product, and so must be seen to be dealt with effectively.

This is a subjective analysis reflecting attitudes and beliefs of those doing the assessment but it does provide a firmer base to consider short- and long-term changes that are needed.

5.3.2 SUMMARY OF THE PRIORITIZING PROCESS

1. ALL risks identified are collated to reduce overlap and repetition, and to establish a list of Primary Risk Factors.
2. Risk factors identified are listed in descending order according to the total score allocated.
3. Plot scores on grid or spreadsheet.
4. Decide criteria for considering urgency of actions required.
5. Identify where urgent action is required, such as clusters of risk factors found in bottom right-hand boxes of grids.
6. Consider what actions can be taken quickly to alleviate the potential harm, whether low or high scoring factors.
7. Identify long-term actions needed, within acceptable timescales that do not further jeopardize the protection of people, property, and the business itself.
8. Consider cost implications, ensuring that they are not used as an excuse to avoid taking any action where it is clearly required to safeguard the future of the firm.
9. Review risk factors and re-evaluate based on potential results of planned actions
10. Consider residual risk, and hazards that will remain despite planned actions. Note which elements are inherently hazardous and can only be contained by a range of control measures.
11. Define "acceptable" or "tolerable" risk and consider the level at which the organization accepts such risks exist, and that sufficient action has been taken "as far as reasonably practicable" in the circumstances.
12. Identify which factors can be reduced to virtually eliminate the perceived risks, for example, through substitution of materials, changed processes and procedures, or transfer of liability.
13. Ensure that any planned changes do not introduce further or new risks to the operation of the business.

5.4 SELF-REFLECTION QUESTIONS

- Was it easier to identify physical, behavioural, or organizational controls?
- Were there any of the possible controls listed that you had not considered before? If yes, what were they and will you incorporate them in the future?
- Were you able to answer the question posed in section 5.2 about how effective existing controls are or did you find some not working as they should?
- Was it easier to use three categories or five to assess the risks?

6 Case Study Examples of Hazards and Risks

6.1 FOUR CASE STUDY EXAMPLES OF POTENTIAL RISK

The featured case studies might highlight some points you have not considered before, even if it is not directly the same business sector you operate in.

To illustrate the way different types of business might apply the suggested approach to identify potential risks, four case studies are considered in this chapter.

There are two service-based industries that are typical of today's provision. These are:

- **Case Study 1: Travel and Hospitality**
- **Case Study 2: Call Centre**

There are also two production-based industries, which are a crucial part of industry around the world. These are:

- **Case Study 3: Food production** and/or Processing, whether as small-scale units or larger industrial plants
- **Case Study 4: Engineering** and manufacture

The potential risks to each of these types of business are considered against the 10 Principles, risk ratings are suggested against each of these elements, and a chart is drawn up to highlight where the priorities for action appear.

6.2 CASE STUDY 1: TRAVEL AND HOSPITALITY

6.2.1 INITIAL THOUGHTS ABOUT LIKELY RISK FACTORS

Travel, tourism, and hospitality sectors were the hardest hit by the global pandemic in 2019–2022 and are still recovering at the time of writing (2023) this book. In recent years, the sector was seen as an example of sound risk management at the level of

DOI: 10.1201/9781003405641-9

holiday provision, dealing with customers, and transferring a great deal of potential liability through insurance cover.

However, the pandemic was an unforeseen external pressure resulting in risks that proved to be uncontrollable for many in the industry. Major risks are associated with competitive issues and finance due to cancellation policies, lack of access to flights and accommodation worldwide, different regulations imposed by popular resorts, and indeed the length of time these conditions remained (Jeynes, 2022). People and protection elements are a vital part of the industry, whether this is for staff, customers, or service providers.

6.2.2   RISK FACTORS GROUP A: PREMISES/PRODUCT/PURCHASING

Issues identified include:

- **Premises** – location is a key element of travel plans and holiday destinations, whether overseas or in the home country. For visitor attractions, resorts, and accommodation, extra cleaning was needed and continues to be an expectation by visitors, particularly in public areas. There are still some restricted regions for travel, including those affected by global conflicts, presenting risks for travellers and those in resort.
- **Product** – visitor attractions have seen numbers of bookings reduced, and many still offer a restricted version of their programme. For example, many cruise ships are limiting the features frequent customers expect to see.
- **Purchasing** – risks associated with pre-ordering tickets/bookings then having to cancel or find places are no longer available. Transport delays, strike action, are adding further risks to manage.

6.2.3   RISK FACTORS GROUP B: PEOPLE/PERFORMANCE/PROTECTION

Issues identified include:

- **People** – a crucial element of the industry, both in face-to-face situations and as hosts or intermediaries. There is a growing risk related to frustration and anger of customers where plans have had to be halted. Staff illness, shortages, and high staff turnover are traditionally associated with hospitality sectors. There is also a long-term issue related to younger age and experience of staff, and their additional training needs.
- **Performance** – delays in transport facilities are outside of the booking intermediaries. Negative reviews and social media feedback, as well as the role of critics and "influencers", present a growing risk for all sectors of the industry.
- **Protection** – during the pandemic, this related to vaccination and certification protocols, for staff and customers, and access to suitable PPE. These risks have reduced although there are still residual risks related to violence and abuse towards staff, particularly in face-to-face situations.

6.2.4   RISK FACTORS GROUP C: PROCESS AND PROCEDURES

Issues identified include:

- **Process** – greater reliance on computerised systems for bookings and payments plus the need to balance customer preferences with what the provider needs. Recent research showed that for the mature travel market, online sources were used to explore options for travel and holidays, but there is still a significant proportion of people who prefer to do the actual booking with a local agent.
- **Procedures** – health certification and visa applications present the greatest risks related to procedures, as does stress for staff and customers when trying to check in at airport terminals when there are unavoidable delays in service. A significant issue relates to compensations/refunds/future vouchers where booked trips are cancelled.

6.2.5   RISK FACTORS GROUP D: PLANNING/POLICY

Issues identified include:

- **Planning** – the industry is still trying to overcome long-term risks associated with the pandemic (which has had a significant impact that many in the industry have found it impossible to recover from). Until 2020, there was real growth and travel was a buoyant sector with many long-term investment plans. Many of these were put on hold or dropped completely. Closures and limited capacity of some providers, including town-centre restaurants or bars, also impact the intermediary's planning activities.
- **Policy** – many have had to rethink financial policies and those related to competition, expansion, and employment. Environmental impact of travel and tourism has also resulted in new approaches to many policy targets.

6.2.6   POTENTIAL IMPACT OF RISK FACTORS IDENTIFIED

Table 6.1 and Figure 6.1 compare a scoring system against five of these risk factors.

6.3   CASE STUDY 2: CALL CENTRE

6.3.1   INITIAL THOUGHTS ABOUT LIKELY RISK FACTORS

A call centre in this case study refers to a service providing a direct interface between customer and supplier of products or services, though obviously at a distance – sometimes completely devolved to another region or country. The most important issues are likely to be related to the Product/People/Process, closely followed by Premises/Protection. The crucial factor is recognition that the philosophy of a call centre is based on providing a human interface with callers, rather than computer-generated responses; therefore, staff members are a vital component of the system. That is,

TABLE 6.1
Scores 1–10 Depending on the Potential Level of Negative Impact

Risk Factor	Who Affected	Severity	Likelihood	Immediacy	Recovery Possible	Cost Implications
Cancellation cost	7	10	7	7	7	10
Changed demand	6	6	8	4	5	7
Booking systems	8	7	8	7	6	8
Places to stay	5	6	6	8	8	9

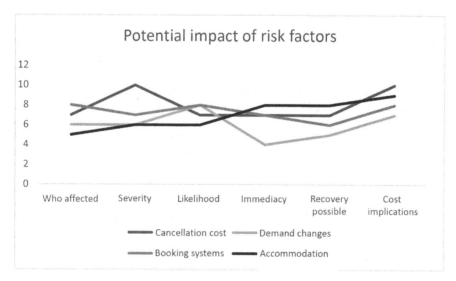

FIGURE 6.1   Travel and hospitality: potential impact of risk factors.

after all, what the client is paying for – humans using technology to support this philosophy.

6.3.2   Risk Factors Group A: Premises/Product/Purchasing

Issues identified include:

- **Premises** – location is less important to client firms or customers as direct access is not necessary, and therefore can be in less expensive area of a country. Rural area has less satisfactory transport infrastructure in place, making it expensive and/or difficult for workers to get there; rising fuel costs add to this problem;

size and capacity of premises are an issue, given the density of equipment needed and danger of becoming too cramped for comfortable working; work environment and conditions are important to overcome the perception of call centres being compared with the old industrial "sweat shops".

- **Product** – service is provided for a third party so removed from client; there is little choice on responses required by operatives, in some cases every word is specified; need to ensure sufficient knowledge of client's product or service; limited capacity available for responses per operative per shift.
- **Purchasing** – not initially considered a significant problem area, mainly related to access to training and other external facilities; equipment must be robust/ adjustable (especially chairs, keyboard, and screen)/suitable for intensive use/ needs good user interface and screen image.

6.3.3   RISK FACTORS GROUP B: PEOPLE/PERFORMANCE/PROTECTION

Issues identified include:

- **People** – high staff turnover; recruitment difficult in some locations; skills needed including ability to speak clearly; training required for new staff, including use of equipment, and adjusting to fit or reduce screen glare; very limited choice available to individuals in ways they work.
- **Performance** – need to set appropriate targets in agreement with operatives; performance measures generally by volume and type of calls/responses made/ time taken/extra "chat" with callers; volume of customer complaints per month; number of clients serviced and retained per annum; costs including wages plus staff turnover and absence levels.
- **Protection** – mainly health issues such as potential hearing loss and levels of background noise; static posture for long periods and potential for repetitive strain injury (RSI); rest breaks/pace of work/intensive use over long period; adequate light, heat, and ventilation levels must be maintained; workspace ownership a potential issue.

6.3.4   RISK FACTORS GROUP C: PROCESS/PROCEDURES

Issues identified include:

- **Process** – provision of information, advice, and guidance on behalf of client; access to supporting information; convergence of telecommunications and computer facilities into single completely self-contained workstations; job design options limited as standard responses required; sitting in one place for long periods and associated stress factors for operatives.
- **Procedures** – use of Display Screen assessment procedures vital; ensure rest breaks taken; dealing with complaints or abusive calls from customers; need to take notice of staff complaints at an early stage to reduce potential problems; escape procedures in the event of a fire or emergency not clear.

6.3.5 RISK FACTORS GROUP D: PLANNING/POLICY

Issues identified include:

- **Planning** – volume of calls and capacity ratios; number of clients serviced (whether in-house or external clients); shift patterns and ensuring adequate cover for staff absence; replacement and investment planning for equipment; return on investment.
- **Policy** – pricing agreements; payment terms agreed; collection of outstanding debts; wage and benefit agreements; equal opportunities across all sites and levels in the firm; confirmation of commitment to taking staff complaints seriously and ensuring conditions are appropriate for the level of work expected.

6.3.6 POTENTIAL IMPACT OF RISK FACTORS IDENTIFIED

Table 6.2 and Figure 6.2 compare a scoring system against five risk factors identified for this case study.

6.4 CASE STUDY 3: FOOD PRODUCTION AND PROCESSING

6.4.1 INITIAL THOUGHTS ABOUT LIKELY RISK FACTORS

With food production, a major concern is about delivery of consistent standards from a more variable base of foodstuffs that can be adversely affected at short notice. The industry is highly regulated, so legislative requirements are a prime issue, and maintenance of adequate control systems. Issues are therefore Purchasing/Product/Process closely followed by Procedures/Protection/Premises. Throughout, People and Planning are also important elements to consider.

TABLE 6.2
Scores 1–10 Depending on the Level of Negative Impact

Risk Factor	Who Affected	Severity of Impact	Likelihood of Impact	Immediacy	Recovery Possible	Cost Implications
Capacity of premises	6	6	6	7	6	8
Location of premises						
Purchasing right equipment	8	7	8	6	5	8
Protecting workers	8	9	7	6	8	6
Use of workstations	8	7	7	8	6	6

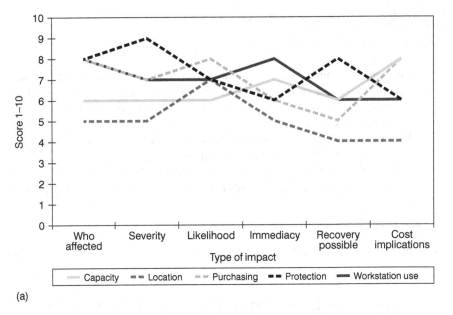

(a)

FIGURE 6.2   Call centres: potential impact of risk factors.

6.4.2 RISK FACTORS GROUP A: PREMISES/PRODUCT/PURCHASING

Issue identified include:

- **Premises** – location and local amenities, particularly problem of unreliable power supply in certain extreme weather conditions; capacity of storage and distribution facilities on site; need for cold storage facilities and specialist delivery vehicles; safe storage of fleet on site and in transit; increasing fuel costs, but good local access to major transport routes.
- **Product** – labelling and packaging requirements considerable, with growing base of information needed to be included for consumers; conflicting scientific evidence makes it difficult to ensure consumer confidence (e.g. on genetically modified (GM) components); need to include guidance on use as well as contents, to reduce potential for litigation claims; need to be able to change range quickly in response to national emergencies (such as bovine spongiform encephalopathy (BSE) or Foot & Mouth crisis), as well as change in fashion trends.
- **Purchasing** – traceability of supplies; consistent access to correct standard of supplies; need to ensure ethical supply where possible; specification standards and tests for contamination; impact of seasonal/climate change on growing and production schedules.

6.4.3 Risk Factors Group B: People/Performance/Protection

Issues identified include:

- **People** – high staff turnover at lower operative levels; limited skill base available locally as rural area rather than urban; some problems with literacy levels (non-English-speaking workers); need to provide wide range of training internally; skills of drivers are important given regular distances travelled for delivery.
- **Performance** – constant need to ensure quality systems adhered to; volume and cost of scrap produced; amount of production down-time or delays in deliveries monitored; processing time and volume targets met monitored (recent problems with old machine awaiting replacement); staff/costs ratios; staff turnover rates and cost of recruitment.
- **Protection** – extremes of heat and cold in process; noise levels for some parts of process; need to maintain adequate ventilation and extraction systems; flow of dust and potential for explosion; protection of consumers with information on contents (allergies); quality assurance scheme for major customers; food and hygiene regulations strictly adhered to as well as H&S (use of Hazard Analysis and Critical Control Point (HACCP) recording system in the UK); use of PPE where appropriate.

6.4.4 Risk Factors Group C: Process/Procedures

Issues identified include:

- **Process** – high use of power, so potential negative impact of climate change levy; significant levels of waste produced naturally; need to maintain machines in constant use; use of high-pressure systems; combination of water and electricity in process, so increased potential for electrocution; capacity of existing machinery and need to update with more computer-controlled, cleaner more efficient versions.
- **Procedures** – usage of water during processing; levels and type of waste produced; waste storage/collection/disposal; potential leakage of contaminated substances/waste; monitoring driver hours and skills/training; evacuation procedures in the event of emergency given the variety of ethnic groups working on site; need to maintain quality and environmental management system standards fully.

6.4.5 Risk Factors Group D: Planning/Policy

Issues identified include:

- **Planning** – main problems associated with reliance on a major customer (Pareto's 80–20 balance) and risks associated with squeezing margins then cutting orders; increased use of e-commerce facilities planned; investment

TABLE 6.3
Scores 1–10 Depending on the Potential Level of Negative Impact

Risk Factor	Who Affected	Severity of Impact	Likelihood of Impact	Immediacy	Recovery Possible	Cost Implications
Labelling and guidance	5	5	6	5	5	8
Location of premises	7	8	6	5	7	8
Purchasing supply traceability	5	8	8	8	8	6
Power usage levels	6	7	5	4	7	8
Local skill base	7	8	7	6	7	7

plans and expected returns for machinery replacement programme; contingency planning required in short term to ensure supplies from new sources (given restrictions of Foot and Mouth outbreak in the UK and elsewhere).
- **Policy** – recruitment based on non-discrimination and equality of opportunity; problems associated with identifying illegal immigrant applicants, given current rural location; minimum wage guidelines followed, although this would represent potential problem if levels raised significantly in future; cash flow and late payment policies in place, using factoring services where necessary.

6.4.6 POTENTIAL IMPACT OF RISK FACTORS IDENTIFIED

Table 6.3 and Figure 6.3 compare a scoring system against five risk factors identified for this case study.

6.5 CASE STUDY 4: ENGINEERING AND MANUFACTURE

6.5.1 INITIAL THOUGHTS ABOUT LIKELY RISK FACTORS

As with the earlier case study on food processing, this is primarily concerned about consistency of supply to customers, although there is likely to be less variation in the quality of raw materials.

Environmental and competitive issues are of concern in this context, as their potential negative impact can be considerable. The important factors are Purchasing/Product/Process/Procedures, closely followed by Premises/People/Planning, although Performance is of course a vital element too.

6.5.2   RISK FACTORS GROUP A: PREMISES/PRODUCT/PURCHASING

Issues identified include:

- **Premises** – capacity and layout; storage for large-scale raw materials/ work in progress/finished goods; access for deliveries and vehicles used on-site; security systems for restricting access to visitors (especially those delivering goods); distance from main transport routes, particularly for exports and materials sourced from outside the UK.
- **Product** – need for consistency of quality standards to customer specification; increased cost of packaging materials and regulatory requirements; minimizing use of packaging to only essential protection; major concern about producer responsibility for disposal of obsolete manufactured goods from customer; need a mix of long-term and short-term design specifications work to optimize the use of machinery.
- **Purchasing** – increased cost and reduced supply of raw materials; use of internet to find new, renewable sources of materials for some work; need to regularly update computer-assisted machinery.

6.5.3   RISK FACTORS GROUP B: PEOPLE/PERFORMANCE/PROTECTION

Issues identified include:

- **People** – fewer tasks requiring non-skilled labour and more computer literacy skills needed; local skill base reasonable, but training on specific machines still

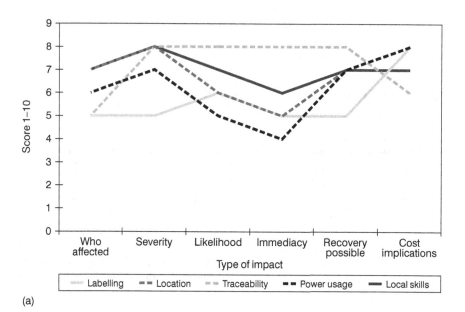

(a)

FIGURE 6.3   Food production: potential impact of risk factors.

required; older average age of workforce – not a problem at present but will become one in around 5+ years; need to bring manager-level skills in from outside for some parts of business.

- **Performance** – client expectations of adherence to quality and environment management system standards (MSS); regular review for accreditation to ISO standards; individual performance targets set and monitored annually, up to senior management level; refresher training provided when possible, but in-house training expected to pass on relevant skills to production workers; returns on investment programme (currently year 2 of 5) monitored; insurance claims and premium reviewed regularly to ensure adequate risk control systems in place.
- **Protection** – traditional areas of health and safety protection still important, with increased emphasis on the use of VDUs; security of design and manufacturing process information; need to ensure existing procedures are in place and working adequately (update risk assessments).

6.5.4   RISK FACTORS GROUP C: PROCESS/PROCEDURES

Issues identified include:

- **Process** – capability of existing plant and equipment to meet current and future client specifications which are becoming more sophisticated; high use of power accompanied by a need to maintain cleaner work environment, so replacement of ventilation system required.
- **Procedures** – emergency procedures for evacuation of plant need updating, particularly for potential hazardous chemical spillage and widespread impact on local community; safety procedures for use of heavy equipment; regular noise level monitoring in place; storage and handling of scrap, plus specialist disposal facilities used.

6.5.5   RISK FACTORS GROUP D: PLANNING/POLICY

Issues identified include:

- **Planning** – need to ensure adequate communication between planning/ production/sales/purchasing departments to avoid last minute scheduling changes (and subsequent costs of scrap and changeover time); continued programme of replacing equipment; skills and recruitment planning needed to match planned changes to production profile.
- **Policy** – equal opportunities but note success of recruiting in non-traditional areas still limited; need to consider rehabilitation policy to retain skills; confirm, in consultation with all interested parties, policy of emphasis on increasing high-specification products that are high-value, rather than high-volume low-value production cycles.

6.5.6 POTENTIAL IMPACT OF RISK FACTORS IDENTIFIED

Table 6.4 and Figure 6.4 compare a scoring system against five risk factors identified for this case study.

TABLE 6.4
Scores 1–10 Depending on the Potential Level of Negative Impact

Risk Factor	Who Affected	Severity of Impact	Likelihood of Impact	Immediacy	Recovery Possible	Cost Implications
Use of safer materials	5	6	7	6	5	7
Quality failures	6	7	6	6	7	9
Disposal of hazardous waste	5	8	6	6	8	8
Security vehicles	6	8	7	7	6	9
Customer complaints	3	5	5	5	6	7

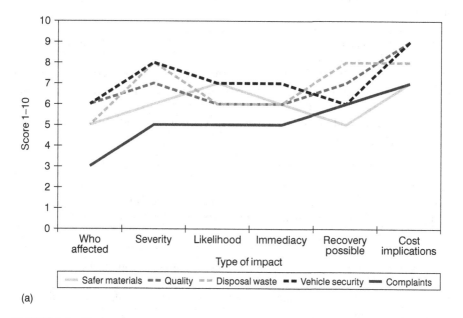

(a)

FIGURE 6.4 Engineering: potential impact of risk factors.

6.6   STRATEGIC OPTIONS FOR CASE STUDY FIRMS

Performance can often become a negative risk factor as it potentially introduces additional criteria that impact on existing factors. On the other hand, it is an important management tool to identify and eliminate unnecessary or uneconomic elements of the business operations.

Earlier sections of the book considered strategic questions in more detail, but it is worth noting here some points that will need closer reflection by firms such as those featured in the four case studies. These will vary depending on geographic location, size, and age of business, and new or proposed changes to legislation, but they are generally seen as typical issues facing these sectors wherever they are located.

6.6.1   Travel and Hospitality

Recovering from a long period of industry downturn due to global conditions is the major issue for those in this sector.

- Cancellation of bookings impacts on financial stability as well as customer dissatisfaction. Insurance cover is not sufficient to spread all the risks.
- Planning for the future is more difficult as global situations change and patterns of bookings by customers have altered significantly.

6.6.2   Call Centre Facilities

Employment and legislative strategic issues seem to be the main ones in this context, including

- Recognition and commitment at the most senior level to valuing the crucial role played by people in the organization.
- Combining Policy/Planning/Performance elements so that everyone is involved in designing the process to improve job satisfaction, reduce potential risks from poor morale/high staff absence/high staff turnover.
- Involvement and consultation used to maintain loyalty and commitment of staff for optimum performance.

6.6.3   Food Production and Processing

There is considerable overlap between different strategic issues in this instance, as competitive, legislative, and financial issues are of concern, such as:

- Establishing and monitoring systems to ensure consistency, quality, and traceability of supplies for the future, including alternative options in unforeseen or emergency situations (such as BSE/Foot & Mouth/flooding).
- A review of risks associated with location, particularly distribution channels and methods.

6.6.4 ENGINEERING AND MANUFACTURE

Competitive and legislative issues are of concern for manufacturing industries at a strategic level, particularly

- The choice of a company-wide approach to Quality/Environment/Health and Safety issues irrespective of differences between sites, local conditions, and legislative requirements, leading to consistency of approach/wider recognition by global customer base/plus standard provision of product or service.
- Commitment to use of sustainable, ethical supplier base for goods and materials.

7 Management Strategies

7.1 STRATEGIES FOR MANAGING THE RISKS

A comprehensive analysis of risks to the business is a vital step in confirming your assumptions and gaining a full picture of the potential extent of harm if action is not taken. However, this has only been an auditing exercise up to this point and so does not demonstrate that risks are being managed effectively or otherwise.

In Chapter 1, it was noted that the 10 Principles are intended to provide a range of prompts to ensure overall coverage of business activities without stressing the importance of one area over another. Clearly a useful model or tool to work with, but in practice these boundaries are not so easily separated out from each other.

While some elements are "operational" rather than "strategic", they are just as important in the context of managing risks to the business. As shown in Figure 1.2 in Chapter 1, all elements of the four central groups of **Physical Properties/People/Process** and **Management issues** feed directly into the **Planning** process and thus into **Policy-making** decisions. So far, these two elements of Planning and Policy have only been discussed in relation to potential risk factors rather than as elements of the management process itself.

The prioritizing activities carried out at the end of Chapter 5 are part of the planning process. Patchy, inadequate sources of data, and lack of management skills or expertise in-house, are likely to have gained a medium to high risk factor rating and therefore need urgent action. Policymaking risk factors, on the other hand, are more likely to have attracted a lower rating on the basis that changes to policy will be far-reaching and long-term rather than needing urgent, immediate attention.

7.1.1 PLANNING

The priority rating for risk factors that need your attention has helped to identify future actions. It does not necessarily represent a set of targets or objectives for the business overall. Depending on the size and structure of the organization, the range of activities needed to reduce or eliminate the risks could well be substantial. Planning activities need to acknowledge other factors to be effective and reach the targets set. These include questions such as actions required/where/by whom/by when, and resources needed.

DOI: 10.1201/9781003405641-10

Actions required are likely to include:

- More in-depth, technical analysis of potential risk factors identified, such as noise levels or air contamination, possibly on a regular surveillance basis over a specified period
- Replacing existing machinery or equipment with less hazardous operational features
- Sourcing new or existing suppliers to find alternative, safer products or materials
- Recruitment of staff or outsourcing some procedures to upgrade skills, knowledge base, and expertise; provision of training for particular groups of workers
- Sale, purchase, or renovation of premises and plant
- Relocation of some business activities to reduce potential risks from over or under capacity utilization
- Evaluation of alternative sources of finance
- Establishing more effective, widespread consultation opportunities that include all relevant groups of workers.

Where, and which business units are directly affected?

- Current and proposed location of business activities, facilities, staff including those working from a home base
- Where to access technical expertise on short-term and long-term basis.

Who will be involved in the activities required?

- Allocate responsibility to make sure actions are taken as planned.
- Devolve authority to ensure decisions are made and actions are taken.
- Arrange for results of activities to be monitored effectively, agreeing criteria for measurement and timescales for each stage.
- Make sure everyone is involved directly in discussions, informed of outcomes and decisions made, and given relevant feedback on progress.
- Confirm the skills or expertise required to carry out the set tasks competently, and that those with responsibility for actions possess such skills or expertise.

When are activities scheduled to take place?

- Agree urgency of actions required, and specify short- or long-term timescales.
- Identify milestones and target dates for monitoring/review/evaluation of results.
- Budget to fit with investment plans, cash flow patterns.
- Make sure that these fit with legislative requirements.

It is vital to ensure that adequate resources are available to enable successful outcomes.

- Allocate sufficient funding to ensure the best fit with improvements required, available at the right time to ensure optimum results.
- Ensure staff have time for extra roles and responsibilities allocated.
- Establish procedures to provide access to data, remembering it needs to be timely, accurate, valid, and relevant for the tasks set.

Although this is very detailed and needs considerable thought to get the most out of the whole risk assessment and evaluation process, it forms the basis of future decision-making and actions. It will now be more about revisiting and monitoring than having to redo it all from scratch. So, a positive result!

However, note that there is no set format to follow, it just needs to be in the format that works best for your business. All the plan(s) of action must be clear to those directly involved and to observers, "transparency" being a key word in this context. Commitment to the planned actions is vital, right from the earliest stages, and is most likely when all those affected by the decisions feel that they have had an input to the decision-making process.

This is true at all levels of the organization, and transparency of management decisions is increasingly seen as a significant factor in how successful an organization is judged by stakeholder groups. It is not just in the context of health and safety that consultation with workers or their representatives is expected and enshrined in regulations. Everyone expects to know more about the firm's activities.

It is much easier to measure and evaluate outcomes of planned activities, and to judge whether objectives and targets have been met, if these are clearly set out at the earliest stages. Such targets or objectives should not be confused with the overall aims, which are likely to be much broader in content and context. For instance, the aim may be to improve the skill levels of staff in a specific division, but to be able to judge if this has been achieved, more detailed objectives need to be stated, such as:

- By the end of 6 weeks a training needs analysis (TNA) will have been completed for all specified workers, and gaps in individual skills or competences identified.
- A relevant training plan will be prepared for every individual, and learning targets set.
- By the end of 3 months appropriate training providers will have been identified and a 12-month training programme agreed.

Although some of the risk factors may have been given a very low rating, considered to be comparatively trivial or low-level risks, a review of the situation must be carried out within a reasonable time frame – probably annually – to confirm that conditions have not materially altered. Changes introduced must be checked to ensure they have not had a negative impact on existing conditions, and that existing controls are indeed adequate and sufficient.

In high staff-turnover environments, introduction of new workers to a job or department can increase the risk factor rating even if all other conditions remain the same.

Factors we considered earlier when deciding priority ratings against individual risks are just as important in the subsequent planning stage. To reiterate, these included:

- Severity – the extent of harm, injury, or damage resulting from exposure to the risk, and of not acting to reduce existing risks; minor or small-Scale damage, localized impact, inconvenient loss, compared with severe, irreversible damage or injury, rapid and wide-ranging loss
- Who will be affected – number of individuals that will be affected, at departmental, business unit, or company-wide; impact on wider community/ environment
- Immediacy – likelihood it will happen and possible time frame, inevitable in near future, or much longer before the impact materializes
- Disruption – does the risk represent significant disruption to business activities, either in the short or in the long term, perhaps just a minor "blip" or a fundamental change to the way the business is organized?
- Ability to recover – does the organization possess the necessary internal skills to be able to deal with the risk? what are the strengths and weakness that will directly impact the ability to recover? will it be able to recover at all?
- Cost implications of getting it wrong – increasing reliance on litigation by customers, clients, and others; the impact on public image, shareholders' views, and on the business owner themself; financial cost to the business.

By this stage, we have covered the fifth principle of the risk assessment process:

monitor and re-evaluate after appropriate time scales, and when circumstances/ materials/processes change.

7.1.2 RANGE OF STRATEGIC APPROACHES FOR DEALING WITH RISKS

It will be clear by now that the approach taken in this book is aimed at developing a holistic, interdepartmental system to reduce the negative impact of functional boundaries, and the lack of effective communication often associated with such barriers. For micro and very small firms this may not be an issue, depending on how we defined a "small firm" earlier.

In addition, the emphasis is on tackling all potential risks to the business equally in the first instance, to overcome problems associated with financial risks taking precedence over operational-level factors related to health and safety and other legislative requirements.

The approach taken by each business will often stem from the attitudes, beliefs, and to some extent the functional area of expertise of the founder or owner of the firm. They will have stamped their own ethical beliefs on the way the firm is organized, the priorities it gives to different risks facing the firm, and the way internal structures are developed. Reference is often made to the "health and safety culture" within a firm, and the overt methods for protecting workers or others. At a broader level, the internal culture will also have impact on the way responsibility, authority, and blame are levelled at individuals or business units.

In this instance, there is only room to mention these facets of the firm briefly, while acknowledging there may be fundamental problems associated with the need to change internal attitudes and beliefs to satisfactorily manage all the risks.
Other approaches to consider include:

- The use of industry standards and "benchmarking" to judge performance
- Formal Management System Standards (MSS), such as the ISO or BSI ranges
- Continuous improvement systems recognized in your state/region
- Financial strategies.

Use of Industry Standards

Some industries have identified "good practice" over many years and produce guidance standards for use by sector players, so you will already be familiar with them. They are generally voluntary, mandatory in some instances, and produced by those working or experienced in the sector to ensure they reflect the real conditions in the workplace. If you need a license to operate, the standards will generally form the basis of license applications, to demonstrate awareness of and commitment to good practice.

Targeted at specific industries, they are often developed from the operational level, with less emphasis on strategic or management issues as these obviously depend on the type and structure of the firm. In this case, some of the risk factors may be missed or assumed to be dealt with differently at senior management level.

Benchmarking, where a firm's performance is measured against other players in the industry, has grown over recent years. Generally associated with larger organizations, there may be value in smaller or medium-sized firms checking their own progress against others to help them develop future strategy for growth. The process includes some form of auditing or reviewing existing conditions to identify gaps and set targets for improvement.

Benchmarking tools are often considered to be motivating for workers, provided everyone is clear about what the intended outcomes are. Many business support agencies have developed useful tools to carry out such evaluations, as have regulatory bodies in relation to health and safety performance. There will be additional guidance available online from your own national governing bodies, so it is worth searching for these according to your industry sector.

Use of Formal Management System Standards

Many larger organizations, certainly in manufacturing and some business support areas, use formally structured MSS that have been developed by standards bodies, with input from relevant industry representatives. Apart from standards related to specific types of activities, these MSS are intended to be generic, and usable by a wide range of business enterprises. The most commonly used ones are for Quality (ISO 9000); Environment (ISO 14000); and Guidance for managing Occupational Health & Safety (OHSAS 18001).

They provide a structure for evaluating the organization at various stages, setting parameters for production/performance, and for establishing a system to monitor, measure, and record performance at these specified stages. Each has until now taken a slightly different approach to structuring the system, as shown below.

- Quality MSS – developed to demonstrate consistency, achieve and maintain a specified standard of performance, they are aimed at achieving customer satisfaction with the product or service.
- Environmental MSS – intended to help firms develop objectives and policies to ensure environmental aspects of the business are managed effectively, within relevant regulatory requirements. This includes reference to wider environmental protection related to the impact of operations on pollution, emissions, and global warming.
- OH&S standards – the original British Standard BS 8800 was a guidance document rather than a formal MSS. The OHSAS 18001:1999 specification produced later incorporated the BS 8800 pattern and is now recognized as one of the major international standards.

Each of these systems has advantages, leading the organization into developing management strategies that aim to identify and control risks. There has been pressure for all three elements of quality, environment, and H&S to be integrated into a single MSS.

Of greater concern is the pressure for external certification of these systems, and the inevitable additional costs to the business that this entails. Perhaps an integrated approach to managing all the diverse risks to the business is more appropriate than an "integrated management system standard".

Continuous Improvement Schemes

There are many models and schemes to help firms develop their internal management systems. It is likely that even the best-run organizations will have some elements of the business that could be improved. The notion of aiming for better performance, rather than becoming complacent once a specified level has been reached, is a valuable one.

Financial Strategies

The cost implications of taking a proactive approach to identifying and managing all these risks can be considerable and the reason why firms are reluctant to start evaluating risks, preferring to ignore them and hope nothing happens to force them to take urgent action.

There are likely to be costs associated with some of the actions required to reduce or eliminate risks, but these must also be considered alongside the potentially higher costs of trying to deal with the damage that could result from non-action. Therefore, some form of cost–benefit analysis needs to be carried out to support and justify decisions made.

The use of formal management systems such as those outlined earlier represent a cost and this may relate to the type of industry sector rather than the size of the firm. We have to recognize increasing public pressure for firms to take an ethical, sustainable approach to operating their business whatever their size.

Financial strategies are likely to include reference to:

- Divestment or acquisition of companies, business units, or plant
- Developing in-house capabilities or outsourcing stages of production
- Investment programmes and timescales
- Use hire purchase or leasing arrangements rather than outright purchase
- Use of credit reference agencies and status reports for clients/suppliers
- Use of factoring and discounting for bad debt
- Spreading risks via insurance cover.

The need to use such strategies should be clearer if the whole range of risks has been considered, and impact assessments are made for proposed actions. These choices are made from a base of significant experience in financial management in large organizations, but for smaller operations such expertise may not be readily available.

7.2 STAKEHOLDERS

We have talked about controlling risks in a practical sense, spending a lot of time coming to grips with the full range of risks facing the business. At the strategic level, the issue is more complex than just eliminating or reducing the risk as much as possible to control potential loss. Measures to avoid risk involve a range of controls.

There is also the question of how to finance potential losses from remaining risks once everything possible has been done to control them. Small-scale events and losses are frequently experienced by firms, but the scale of costs is not fully realized or is overlooked. On the other hand, experience of a serious event is rare in most firms, so the impact is often underestimated. If the risk relates to personal injury or death, it must be reduced to as low a level as possible, but for other types of risk you might decide to accept the potential loss rather than utilize more resources to reduce it further.

The cost–benefit analysis referred to earlier is often problematic in relation to health, safety, or environmental risks. This is largely due to difficulties in quantifying the potential loss and benefits in financial terms, plus the time lag between expenditure and resulting reduction in loss. This will be determined by the type of enterprise, industry sector, level of contact with the end user of the product or service, and the various groups of people who have an interest in the firm – the stakeholders. For our purposes, stakeholders include:

- Workers and employees; worker representatives at all levels, part of recognized unions or not, who seek to maintain protection for workers as well as the firm.
- Senior management and/or Board of Directors, who hold ultimate responsibility for losses from mismanagement of risks plus personal liability for breaches of health and safety law as well as corporate liability.
- The owner (as an individual or holding company) who will want to see the optimum use of all resources, and potential risks managed effectively to reduce losses.
- Shareholders who have invested resources into the business expect potential losses to be as low as possible to gain the expected return from their investment.

- Banks and financial institutions who need to be sure that borrowing is correctly geared relative to turnover, profit and investments, and repayment is secured.
- Enforcement authorities, local and national, who expect to see evidence that regulatory requirements are adhered to, proper records kept, and some form of management system exists to safeguard individuals, the public, and the organization.
- Insurance providers who expect that all risk factors have been identified, adequate controls are in place to eliminate or reduce potential losses, and insurance cover provided is appropriate and sufficient.
- The public, whether as consumers, part of the local community, or wider community, is concerned that risks arising from business activities are adequately controlled.

Perception of risk is an interesting factor in relation to different stakeholder groups, influencing their views on what is considered an "acceptable" level of risk. Some older, traditional industries display a level of acceptance of safety or health risks that is considered totally unacceptable by others. The issue of whether the risk is taken voluntarily – that is, as a matter of choice by the individual – is also a factor, particularly in the grey area between work and lifestyle choices such as smoking or sports activities.

An acceptable balance is very difficult to achieve in the leisure industry, for instance, where customers expect to be able to smoke if they wish, but workers need to be protected from the perceived risks associated with passive smoking. This can be of greater concern when the business operates across different countries that have more/or less stringent regulations and cultural values. The question of control over the situation or choice by the individual is, therefore, a crucial factor of perception and acceptance of risk.

7.3   SPREADING THE RISK

Losses are generally divided into two main groups: consequential loss and direct loss.

a) Consequential losses are more difficult to quantify, are not always immediately apparent, and therefore tend to be overlooked or underestimated. They include lost time and business interruption caused by accident investigations, machinery breakdowns, transport delays, supplier problems, disputes with workers, and external events such as flooding. They may also include other elements such as cash flow difficulties and late payment of debts, loss of records and important data, as well as losses associated with bad publicity and poor public image following a serious event, fines, or penalties.

b) Direct losses are the more obvious ones related to replacing equipment or machinery, repairs to plant and premises, damage or loss of goods, and payment of third-party claims.

Broadly speaking, the risk of direct losses is transferred to an insurer, but consequential losses are uninsurable and retained by the organization. If small-scale losses occur frequently, they are probably retained by the company as increased premiums based on claims record may be disproportionately higher than the value of losses incurred.

In addition, some infrequent severe losses may be uninsurable, such as flood damage or bomb explosions. We have all seen the losses incurred due to an unprecedented pandemic in 2019–2021 as well as significant global conflict.

Speculative risks, where the outcome may be a win or lose situation, are unlikely to be insurable as are losses due to inadequate security measures being in place, such as theft of cash from open tills rather than while stored in a locked safe. Residual risks can be spread through a combination of transfer (to an insurer or other external source) and retention (within the organization), by covering the first part of a claim by an excess payment, or by part-insurance where just a percentage of the risk is covered.

Experience suggests that many firms rely heavily on the insurance route to protect them from losses, mistakenly believing that all the risks are covered including consequential loss, rather than taking appropriate action to avoid or reduce the risk in the first instance. It is important, therefore, that you consider different ways to spread the impact of potential losses, making a realistic appraisal of the real losses likely to occur from residual risks. While some losses associated with legislation or security risk factors are likely to be insurable, this is less likely to be the case in relation to competitive and financial risks or indeed some aspects of employment.

7.4   THE IMPORTANCE OF HAVING REALISTIC POLICIES

In most countries/states around the world, legislation requires at least a policy statement on health and safety in some form. There are different requirements depending on the size of the business and how many people work there. For instance, it is required if just one person is employed in Canada or more than five people in a UK business. A typical list of features in such a policy for a smaller firm might include commitment to:

- Providing a safe and healthy work environment for people
- Ensuring premises are maintained properly, good housekeeping standards are kept, and adequate facilities are provided for workers and others on site
- Producing a product that does not jeopardize the safety and health of others, or the environment
- Purchasing less hazardous raw materials where possible that are healthier, safer, and more environmentally friendly to use
- Identifying hazards and assessing risks to workers and others who may be affected by activities of the firm
- Involving workers directly in discussions about health and safety issues or concerns, to ensure their input and commitment to working together to tackle these issues
- Providing sufficient resources, information, and training to people to ensure they can carry out their duties and fulfil their responsibilities in a healthy and safe manner

- Making sure that procedures intended to safeguard the environment are followed correctly, and that processes are adequately supervised
- Ensuring suitable monitoring and recording systems are in place
- Providing adequate protection for people against damage to health, harm, or injury, resulting from work activities or fire risks
- Ensuring processes are carried out using equipment and machinery that is appropriate, as safe to use as possible, and properly maintained
- Setting targets to reduce accidents and ill health in the workplace
- Regularly reviewing the situation to see whether targets have been met, existing controls are still adequate and in place, or new targets need to be set.

A daunting list for anyone! It is not comprehensive but is sufficient to identify intent on the part of the firm to manage your health and safety responsibilities in a positive way. It can be extended to include other areas such as smoking, use of drugs, bullying, and fire protection. The list broadly follows the 10 Principles elements discussed earlier, although it does, of course, concentrate exclusively on the health and safety risks.

As shown in Figure 1.2 in Chapter 1, the elements referred to as 10 Principles all interact with each other, feeding into the planning and decision-making process and the development of relevant policies. It is a continuous process, requiring regular monitoring/review/re-evaluation to ensure it reflects the dynamic nature of the business, rather than being a one-off activity of limited value to any stakeholder groups.

7.5   WIDER POLICY STATEMENTS

However, in the context of risk management generally, this list is clearly insufficient. It does not cover policies on environmental protection, security of data, or even financial policy. You have now identified, and evaluated in detail, a wide range of risk factors, risk control measures, remedial actions prioritized, and management strategies developed. It is vital that appropriate policies are established to support the agreed strategies, that everyone knows and understands them, and that their effectiveness is monitored over time.

This section highlights further areas where a policy statement is needed to confirm commitment at the most senior level within the organization.

7.5.1   PREMISES

Policy issues related to premises are likely to be about facilities and location, such as:

- Maintain buildings, site, and surrounding area as necessary
- Use local sources of labour where possible, providing adequate relocation packages for workers and their families if needed
- Ensure adequate site access for people with physical disabilities, whether workers or customers, based on the existing physical structure of buildings
- Safeguard the site and surrounding areas from potential harm or damage from fire, pollution, noise, and other recognized hazards.

7.5.2 PRODUCT OR SERVICE

A policy statement related to the product or service also needs to acknowledge concerns of the end-user, so statements will include:

- Provide a safe product or service for customers, with relevant information and adequate packaging.
- Provide environmentally friendly product or service as far as possible with safe disposal of obsolete products or waste.
- Comply with relevant consumer protection legislation and offer advice/information/support through Customer Support service.
- Ensure proper training and supervision of workers to produce a product/service conforming to specification.
- Safeguard access to sensitive data and valuables, whether online, on-site, or in transit.

7.5.3 PURCHASING

This section should include reference to policies about identifying, accessing, and paying for supplies as well as the logistical elements of delivery and storage. Policy statements are therefore likely to refer to the following:

- Ensure prompt payment of bills to ensure consistency of supply.
- Establish systems to check quality specifications for supplies are met, appropriate to the type of goods or services being supplied.
- Ethical purchasing policy based on the use of suppliers that operate within the law and do not endanger the health or safety of their workers or the environment
- Ensure safe storage, handling, and transport of goods.
- Use the most cost-efficient purchasing and delivery methods.
- Seek to find more sustainable sources of raw materials and substitute with less hazardous substances where possible.

7.5.4 PEOPLE

Given such a wide range of legislation related to worker protection, it is vital to fully reflect these responsibilities in your policy statements. This is important as a means to support your aims and goals, not just because the law requires it.

While the detail for such policy statements can be developed from guidelines produced by legislators in your region, the following list suggests issues that need to be covered:

- Make a positive commitment to equality of opportunity in recruitment/training/promotion practices.
- Strive to reflect the local population mix within the workforce, and avoid discrimination on the grounds of gender, race, religion, and age.

- Make adequate provision for the employment of people with limited physical or mental abilities, and have access to a rehabilitation programme to aid a speedy return to work.
- Provide access to relevant training, information, and guidance to ensure people have the necessary skills and competence to carry out their job satisfactorily.
- Follow family-friendly working practices where possible, to offer flexibility and encourage worker loyalty.
- Intolerance of workplace bullying or threatening behaviour, aiming to reduce potential for unnecessary levels of stress.

7.5.5 PERFORMANCE

To ensure transparency and commitment company-wide, it is worth stating the policy related to setting targets, identifying performance criteria, and measuring outcomes in the following areas:

- Agree and define criteria for individual/teamwork targets with performance measures that are relevant, achievable, and measurable.
- Establish targets to reduce levels of waste/scrap; review performance against specified criteria in all areas of business operation on a regular basis and on a timely basis to ensure appropriate remedial action can be taken quickly to limit potential damage.
- Establish financial targets and performance measures for future operations and growth.
- Monitor returns on investments regularly to ensure optimum returns are made within targets set.
- Comply fully with corporate governance requirements, based on transparency, agreement, and commitment at the earliest stages.
- Maintain a positive image and standing within the local community.

7.5.6 PROTECTION

Policies must, of course, reflect the need to protect individuals, communities, property, assets, and the wider environment, so may well include quite diverse elements such as the following:

- Assess health and safety risks to workers, customers, and others to ensure that adequate controls are in place to reduce risks to as low a level as possible.
- Provide necessary PPE where other control measures cannot reduce risks any further.
- Take sufficient security measures to ensure protection of assets.
- Ensure funds are invested prudently and ethically to safeguard the future of the organization.
- Protect data in whatever format, sensitive information on workers or clients, according to the requirements of data protection regulations in the countries you operate in.

- Assess risks and ensure adequate controls are in place to protect the local and wider environment from damage such as emissions, fire, and pollution.
- Make efficient use of energy sources to reduce the impact on climate change.

7.5.7 PROCESS

Many of the previous policy statements will cover aspects of the process, such as "Product" or "Procedures", but the following are also policy positions that need to be specified:

- Ensure plant and equipment is appropriate for the job and are regularly maintained to ensure a safe and healthy operation.
- Replace and upgrade as necessary with safer, cleaner versions of the plant to maintain optimum levels of production.
- Identify storage/transport/distribution economies where possible.

7.5.8 PROCEDURES

Policy statements relative to procedures within your organization will reflect the individual nature of the business, but are likely to include the following:

- Consult with workers and others on issues that affect them in the workplace, providing opportunities for them to elect representatives as necessary.
- Ensure discipline and grievance procedures are developed with the intention of protecting people and ensuring fair treatment.
- Ensure safe systems of work and all procedures designed to safeguard health, safety, and security of workers are adhered to.
- Comply with requirements of any management system standards that are in place.
- Ensure that reporting procedures are correctly followed, whether internally required or for external purposes.
- Comply with the requirements of relevant legislation.

7.5.9 PLANNING

As we have already seen, the planning process combines feedback from several quarters, and policy statements should identify how this is to be managed effectively, for instance:

- Incorporate input from all sectors of the organization where issues will impact their working situation.
- Ensure adequate contingency plans to safeguard workers, business operations, and the wider community.
- Make sure everyone knows what these contingency plans are.

- Establish and maintain workable communication channels throughout the organization.
- Take necessary actions to minimize the potential for fines, penalties, or prosecutions against the organization or any of its component business units.

It is quite likely that you will have additional policy issues included to reflect the unique situation in your organization, but the list above covers the main policy areas that should be addressed. In addition, health and safety policy statements need to include specific details of responsible and competent people. You must also include the name(s) of the most senior director(s) with personal responsibility for ensuring compliance with relevant legislation and policy statements, and that health and safety issues are given due priority in management decisions.

Following these policy statements to show all areas have been considered and statements drawn up to establish them across the company, they must of course be brought to everyone's attention. Ideally, people at all levels of the firm will have been involved in discussions about their content and have therefore demonstrated some commitment to the principles enshrined within them. This commitment must be demonstrated from the most senior levels down, as well as the shop-floor up, to maintain the momentum and ensure cynicism does not creep in.

In the context of this publication, a critical element of the exercise is to ensure that the whole range of risk factors is considered equally, without greater emphasis on one aspect over others. It will be clear by now that the greatest challenge for most businesses is to combine the full range of strategic and operational concerns in such a way that protection against existing and potential future risks is an integral part of the management process.

By now, you will have a great deal of information about the business. Certainly, there will be a comprehensive picture of what actually happens on a day-to-day basis and a realistic range of policy statements designed to eliminate or control risks.

7.6 SELF-REFLECTION QUESTIONS

- Does your planning process include all the factors listed – actions required/ where/by whom/by when/required resources allocated?
- Have you revisited your notes and made sure that everything has been covered.
- Which strategic approaches have you decided to use to help manage the risks? For example, does your industry sector produce its own standards or guidelines?
- Do you have a full range of policy statements that reflect what happens in your business?

8 Risk Management

8.1 MANAGING THE RISKS

As you can see, there is no single right way to manage the myriad of risks facing businesses today, given the considerable changes in working patterns, globalization, and the unprecedented increase in the use of electronic media to support operations since the turn of the 21st century. The 10 Principles approach aims to bring together the most common elements in a way that recognizes the importance of all risk factors to the successful operation of the firm, and cuts across management functional boundaries.

Risk management skills embodied within different functional areas have a crucial role to play in the risk assessment process and therefore must be brought together to make best use of them. Whether you are a small private sector enterprise, a public sector agency, or a charitable trust, the risk factors identified will apply to a greater or lesser extent to all of them. Most safety, environment, and fire protection legislation applies equally to all but the smallest of enterprises. The risk assessments you have carried out so far demonstrate just how far-reaching the potential harm to the whole organization can be if existing controls are ineffective.

We have seen how different definitions of risk management can be. Generally, the point is that hazards are identified and risk factors analysed, existing controls and any new ones are considered alongside decisions about how to make sure they continue to be effective.

8.2 IDENTIFYING THE RISK FACTORS

On reflection, Part 2 of the book is a comprehensive section that considers the risk assessment process in some detail, both for those new to the process and those with experience of assessing risks in just one area. Many of the principles are the same, of course, but hopefully additional points have been raised that may have otherwise been overlooked. It is always useful to review existing risk assessment procedures periodically to ensure they still fulfil their original intentions, and to reassure people of continuing senior-level commitment.

DOI: 10.1201/9781003405641-11

Experience has shown that a site plan is a place to start the hazard identification process, helping those who are already working on-site and those who are bringing a "fresh eye" to the risk assessment process ensure risk factors are not overlooked due to familiarity. It also reinforces the whole-site, inclusive nature of the assessment, and factors other than physical risk factors may become more apparent. For example, joint use or ownership of premises is often accompanied by conflict over responsibilities for losses arising from uncontrolled risks.

Legislative risk factors impact widely relative to premises, as they do in relation to workers and others on site, so this is often the initial focus for firms wishing to spread potential loss through insurance cover. As we have noted earlier, on its own this does not necessarily lead to enough protection.

It is reasonable to assume that the choice of product or service provided is much more within your direct sphere of control, as are the risk factors associated with this choice. However, as this is so closely tied with purchasing decisions, processes used, and protection of people whether workers or customers, the risk factors are extremely far-reaching.

Demographic changes not only have impact on the range and type of goods or services being sought out by customers, but also have impact significantly on the make-up of the workforce in some industries. The opening of a day care centre by a major multinational firm for older relatives of staff, rather than a crèche, reflects the potential problems many workers now face with these additional responsibilities, and is an excellent example of a positive risk management strategy.

Greater willingness on the part of consumers to seek compensation when unhappy with the goods, a rapidly expanding base of legal firms willing to help them, and a substantial base of consumer protection legislation on content/structure/packaging of the product all combine to represent a powerful risk factor for most industry sectors. While these consumers may be better informed than previously, media coverage is not always as positive or helpful as some would like and often fuels concerns about misinformation. This is a critical element for all firms as social media comment is unregulated, far-reaching, and immediate in its impact.

Although there is still debate about the causes of climate change at a global level, pressure will continue to mount for industry to demonstrate that it takes the issue seriously and is aware of environmental concerns that people have. Ethical purchasing is an issue worthy of consideration, though not necessarily clear-cut in its application, as is the need to ensure fair and equitable trading conditions across borders.

Changing work patterns and increased worker protection legislation over recent years both represent potential risk factors for firms, not least in the need for ever more complex systems in place to ensure adequate protection for individuals. Flexible working and non-standard employment contracts make it difficult to track workers, to involve them in meaningful discussions about issues that affect them when they are mainly off-site, and to ensure that they receive sufficient skills or health and safety training.

Flexibility may have been the byword for competitiveness during the 1990s, but the downside was that it also resulted in uncertainty leading to greater stress. More than 30 years on, flexibility is still a competitive advantage though what form it takes is ever evolving.

There is increasing evidence of workers exposed to violence at work, particularly when working in face-to-face situations with the public, and safety while driving for work purposes is taking on greater significance. These issues represent risk factors for the organization as well as the individual, with the attendant potential for loss that must be managed properly and fairly.

As we have seen throughout, a crystal ball would be extremely valuable to help assess potential risks! For instance, risks associated with the use of substances or processes that may, in the future, be deemed to be more harmful than previously thought can lead to claims for damage after a 20 or 30-year time gap. Having said that, it is likely that everyone can think of examples where workers have complained about unacceptable side effects or ill-health while carrying out certain tasks, but legally they are considered "safe" at present.

Finally, the risk factors identified at this stage of the process rely on the collection of valid, relevant data. Historical data such as absence records, accident reports, previous years' figures and returns, all have their place, but it is hoped that the wide range of suggested factors included here has led to the use of a variety of sources to ensure everything is covered.

8.3   EVALUATING THE RISKS

The use of the ten headings, the 10 Principles, is a useful tool for ensuring the full extent of business operations are considered equally, particularly when combined with the human resource, legislative, security, competitive, and financial risk factors. For those working in some of these areas every day, the risk factors may be glaringly obvious, but as it is likely that other people will be taking part in a comprehensive analysis of the whole firm, it is important to state them clearly.

It is worth spending time to make clear what the potential harm, injury, or damage is likely to be from the risk factors identified, and who is most likely to be affected – individuals, stakeholders, or business units. Potentially, there will be tangible losses from some risk factors, particularly those where specific individuals are most likely to be exposed to the hazard. On the other hand, there are also likely to be less-easily defined losses such as public image or credibility in the local community, as well as far-reaching cost implications for access to markets or services in the future.

How immediate the impact or loss will be should also play a role in the equation, given that the impact may be immediate and low-key or indeed be a long-term problem leading to a fundamental change in the way you operate.

The risk rating approaches suggested in Chapters 4 and 5 are typical for this type of assessment activity, requiring a subjective evaluation to some extent. Computerized assessment systems can be quick and easy to use to generate results. In their absence, the approach suggested here will give a comprehensive, consistent picture to help decision-making about prioritizing future actions. A numerical scoring system gives an opportunity for finer gradations between ratings but may become cumbersome if a substantial base of risk factors is being addressed or the scoring range is too great.

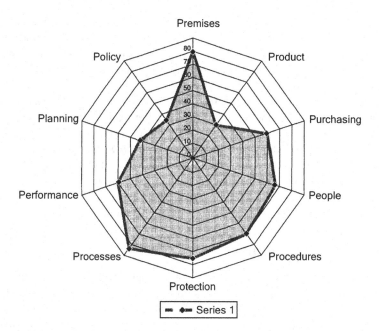

FIGURE 8.1   Total score for risk factors.

In Figure 8.1 each of the ten categories has been plotted in a spider-chart format to illustrate exactly where the biggest risks are concentrated. In this example, the biggest risks seem to be associated with the processes and premises themselves. Clearly, there needs to be some fundamental revision of the operational side of the business and urgent attention to risks identified.

8.4   CONTROLLING THE RISKS

Following our risk assessment process discussed so far, we can move forward to manage the risks in the best way possible. We have considered physical, behavioural, and organizational controls as a means of reducing the negative impact of exposure to hazards or risk factors. These range from decontamination facilities, incentive schemes, and recruitment policies to the use of formal management system standards.

You will already have a diverse range of control measures in place, some more effective than others. Assumptions about existing controls should be questioned to confirm that

a)   They still exist as they were originally intended to.
b)   They are still appropriate for the changing face of risk factors facing your business today.

Procedural controls should certainly be examined, as the fact that they appear in written form as part of contracts or procedure manuals does not necessarily reflect that they are followed. Insurance cover is an interesting point to consider in view of the risks identified, the extent to which existing cover is adequate, and the length of time since premiums and cover were reviewed and reassessed. The issue of insurable rather than uninsurable losses is a crucial one for discussion at this and later stages of the process.

As one of the primary reasons for carrying out risk assessment is to help in deciding priorities for future action, the work undertaken earlier must feed directly into this decision-making process. The most significant risks should emerge as those that require speedy remedial action, and others with a lower rating as less urgent.

However, as noted earlier, even the lowest scoring factors still represent a need for action, and lots of small-scale low-cost changes could represent a more positive impact on other risk factors, such as low motivation or morale of workers fed up with poor local working conditions they have complained about for months.

8.5   MANAGING THE RISKS

The whole purpose of this book is to lead you through the risk assessment process, to a review of controls and identification of gaps in protection against loss or damage. That may be an admirable stage to reach for some organizations where this has not been formally tackled to any great extent. However, it is only part of the story, and in effect demonstrates mismanagement of risks if no further action is taken to eliminate, reduce, or spread the risks and safeguard the interests of all stakeholders in the business.

It is the responsibility of management, whatever the size of the firm, to take appropriate action to minimize the impact of losses on the business and to protect the interests of everyone. Effective corporate governance is a legal requirement in some regions, and although reference is made to directors in this context, it is not only applicable to incorporated enterprises but also to those at senior management levels of all enterprises.

The main principles of any code on corporate governance include the need to:

- Maintain a sound system of internal control to safeguard shareholders' investment and the company's assets
- Conduct a review of internal controls and their effectiveness, including financial, operational, compliance controls, and risk management
- Report findings to shareholders, including explanations of how the principles of this code have been applied.

Guidance [Institute of Chartered Accountants, for example] notes that a risk-based approach "should be incorporated by the company within its normal management and governance processes", rather than as an exercise just to comply with the law.

- In addition, principles behind effective management of health and safety risks should:

- Maximize the well-being and productivity of all people working for an organization
- Stop people getting injured, ill, or killed by work activities
- Improve the organization's reputation in the eyes of customers, competitors, suppliers, other stakeholders, and the wider community.
- Avoid damaging turnover and profitability
- Encourage better relationships with contractors and more effective contracted activities
- Minimize the likelihood of prosecution and consequent penalties.

Clearly, these are all powerful arguments. Irrespective of the legal imperative to comply, all businesses must take on board the need to assess, control, and manage all risks comprehensively in a transparent way that demonstrates their full commitment to the process.

The standard priority for actions to control risks is based on the following order:

- **Elimination** of the process or procedure that presents the hazard
- **Substitution** for a less hazardous material, process, or procedure
- **Restricting access** to the hazard
- **Physical guards or controls** to keep people away from potential harm or injury
- **Procedures** changed to make the process less hazardous
- **Training and supervision** for staff to work safely
- **Personal protective equipment** – the last option after others have been put in place.

There is a useful Management Checklist included in the Appendix to help plan and carry out necessary actions to reduce or eliminate risks identified. There may be other actions that are specific to your organization, of course, so you can add to, or amend, the checklist as necessary.

9 Conclusions

Congratulations on getting this far! We know this has been a long journey through the risk assessment process, especially if this is a completely new venture for you and your business. We have covered a lot of ground but inevitably there will be some aspects that do not easily relate to your business, and others that you will have added along the way.

In addition, everyone in the organization should know what risks exist, the controls in place to limit any damage or injury, and understand the need to make sure procedures and controls are followed.

The 10 Principles approach to risk assessment and subsequent risk management encompasses all the elements needed to demonstrate compliance with any formal code. It involves a holistic evaluation of risk factors, taking positive management action to protect the interests of all parties against potential loss.

The crucial point is that the bulk of the work has now been completed and you are in a much better position to set out your risk management process. As situations change, it will be easier to revisit and amend the existing data than starting again from scratch

9.1 WHAT HAVE WE COVERED SO FAR?

Risk Assessment – what it is, what the process involves, and how important it is. Stages included:

- Identify hazards that could potentially cause harm/injury/damage
- Thinking about what this harm/injury/damage might be
- Who could potentially be affected and how serious would the harm/injury/ damage be to the individual, the company, and the wider environment
- Evaluating the level of risk this represents.

Risk Management – controlling and managing residual risks. Stages included identifying:

- What controls are already in place to reduce or eliminate the risks

DOI: 10.1201/9781003405641-12

- Where are there any gaps in controls
- What extra actions are needed to control residual risk.

Strategies for managing risks, including:

- What actions are required – where/who by/when/resources required
- Planning priorities for the actions required
- Different approaches possible to help manage the risks
- Policy statements related to each of the ten principles and risk factors.

We started with ten risk factors that impact the risk management:

- Physical properties related to the **premises, product,** and **purchasing supplies**
- People elements related to the **people** in the organization, **procedures** they follow, and **protection** measures in place
- Actions including **processes** involved and **performance** measures
- Management issues related to **policy,** strategies, and **planning.**

The risk management principles we have worked through are based on these factors. For us the ten principles of risk management are based on assessing and analysing the situation across all functional areas with input from all those involved with the company:

- Identify hazards
- Identify possible harm/injury/damage likely from exposure to these hazards
- Assess the risks
- Check existing controls
- Identify gaps where more controls are needed
- Decide priorities for actions needed
- Produce policy statements
- Inform all stakeholders
- Monitor effectiveness of changes introduced
- Review regularly to ensure no additional risks have been introduced over time.

9.2   WHAT NEXT?

It is time for action! You have identified all the major hazards and risks in the business and found where gaps in controls exist. Once policy statements are produced, however detailed or brief these need to be for your particular business, and everyone in the organization is notified, it is time to start on the list of actions.

Priorities are based on your assessment, whatever scale you have chosen to use, so a timetable of actions is the next step. Remember that monitoring and review will be an ongoing process to keep you and the business safe in the future.

The self-reflection questions at the end of each chapter will help when putting all the information together. In addition, there are lots of checklists included as appendices to remind you of the main points to look for.

The Bibliography/Reference list includes some useful information sources in various countries. There are also contact details for the author.

Useful Reference Sources and Appendices

RISK MANAGEMENT REFERENCES

Core subjects of Risk Analysis, Society for Risk Analysis, www.sra.org/
Risk Management Framework, US Federal Government guidelines, www.nist.gov/risk-man
agement
11 Principles of Risk Management, www.mybusiness.co.au
7 Principles of Risk Management, www.aipm.com.au
7 Rs of Risk Management ISO 31000, www.iso-consultants.com
6 components of Risk Management, www.worksafe.nt.gov.au
4 Cs of Risk Management, www.unit4.com

PUBLICATIONS OF INTEREST

Jeynes, J (2000) *Practical Health & Safety Management for Small Businesses* published
by Butterworth Heinemann (Elsevier):ISBN 0-7506-4680-2 [see details of revised
version below]
Jeynes, J (2022) *Managing Health & Safety in a Small Business* published by Business Expert
Press, New York: ISBN 978-1-63742-195-6
Jeynes, J (2022) *How a Global Pandemic Changed the Way We Travel* published by Business
Expert Press, New York: ISBN 978-1-63742-301-1
Health & Safety at Work journal, published by Lexis/Nexis hsw@lexisnexis.co.uk, www.health
andsafetyatwork.com
Combined Code of the ICA Turnbull Committee on Corporate Governance (known as the
Turnbull Report) 2000 ICA: ISBN 1-8415201-0-1

GOVERNMENT DEPARTMENTS

UK: Health and Safety Executive (HSE)
  HSE, Redgrave Court, Merton Road, Bootle, Merseyside L20 7HS
  www.hse.gov.uk
  www.gov.uk/health-and-safety-executive
UK: Environmental Management
  www.gov.uk/topic/environmental-management/environmental-risk-management
  www.gov.uk/government/organisations/environment-agency
  www.doeni.gov.uk, Department for Environment Northern Ireland
Australia: www.environment.gov.au
Department of Trade & Industry: www.gov.uk/organization/department-of-trade-and-industry
Other examples of similar sites:
  www.industry.nsw.au, Australia
  www.thedti.gov.za, South Africa
  www.tid.gov.hk, Hong Kong

www.dti.gov.ph, The Philippines
www.census.gov/foreign-trade/balance, USA trade with the UK
www.meti.go.jp, Ministry of Economy, Trade & Industry Japan
UK: Office of Fair Trading
www.gov.uk/en-GB/government/organisations/office-of-fair-trading
UK: Employment law guides, www.employmentlaws.co.uk
  Chartered Managers Institute, www.managers.org.uk
  Institute of Chartered Accountants UK, www.icaew.com
  Institution of Occupational Safety & Health (IOSH), www.iosh.co.uk
  British Medical Association (BMA), www.bma.org.uk
  Occupational Health resources, www.hse.gov.uk/resources/publications/freeleaflets
  British Safety Council, 3 Brindley Place, Birmingham B1 2JB UK,
  www.britsafe.org
  Royal Society for the Prevention of Accidents (ROSPA), www.rospa.com
  The Fire Protection Association, www.thefpa.co.uk
UK: British Standards Institute (BSI), www.bsigroup.com/en-GB
  Quality Standard ISO 9000
  Environmental Standard ISO 14001
  Carbon Footprint Verification Standard ISO 14064-1
  Occupational Health & Safety Standard OHSAS 18001
  Food Safety Standard ISO 22000
  Business Continuity Standard ISO 22301
  Social Responsibility Standard ISO 26000
  Information Security Standard ISO/IEC 27001
  Supply Chain Security Standard ISO 28000
  Energy Management Standard ISO 50001
  Asset Management Standard ISO 55000
  Personal Information Management Standard BS 10012
  Organizational Governance BS ISO 13500
  Risk Management Standard BS ISO 31000
Industry groups such as:
  Federation of Small Businesses, www.fsb.org.uk
  Confederation of British Industry, www.cbi.org.uk
Europe:
  European Small Business Alliance, www.esba-europe.org
  European Small Business Portal, www.ec.europa.eu/small-business/index_en.htm
USA: Small Businesses USA
  https://business.usa.gov, Business USA
  www.sba.gov, Small Business Administration USA
Canada:
  Small Businesses Canada, www.smallbusinessescanada.com
  www.canada.ca/small-businesses

Appendices

APPENDIX 1
DEFINITIONS OF SMALL FIRMS

Definitions of small firms – clearly not easily defined as different sources state different numbers and criteria (based on sources accessed in 2023). Given these often-substantial differences, for our purposes, it may be more appropriate to work on how you perceive your business as a small firm as long as you do not exceed these criteria. For example, it may be more about how you run the business on a day-to-day basis and the administrative systems you have in place to help you.

USA

- Small Business Administration (SBA) source:
  Generally viewed as fewer than 500 employees
  Revenue less than $16.5 million

Definitions range from up to employees in some sectors but between 100 and 1,500 employees in other industry sectors, as shown in the table below:

Industry	Number of Employees	Annual Revenue
Manufacturing/Mining	Up to 500 employees	Not classified
Wholesale	Up to 100 employees	Not classified
Retail and Service sectors	Not classified	Up to $6 million
Construction	Not classified	Up to $28.5 million
Agriculture	Not classified	Up to $0.75 million
Special trades/contractors	Not classified	Up to $12 million

- *Business News Daily* quotes different standards from SBA

Industry	Number of Employees	Annual Revenue
Manufacturing	500–1,500 employees	Not classified
Wholesale	100–250 employees	Not classified
Retail	100–200 employees	$8–$41.5 million
Service sectors	250–1,500 employees	Up to $41.5 million
Construction	Not classified	$16.5–39.5 million
Agriculture & Fisheries	Not classified	$2–30 million

US Chamber of Commerce noted that 99.9% of all businesses – 32.2 million of them – would come into the definitions of a small firm. There were a record 5.4 million new business applications between 1995 and 2021, with 63% of new jobs created by smaller businesses.

In 2022, nearly half of small business owners noted they now work longer hours than they did a year before, and the lack of workers is a significant issue for them.

CANADA

- General view is fewer than 100 employees and between 30,000 and 5 million Canadian dollars
- In July 2022, www.statcan.gc.ca noted that the largest proportion of the labour force worked in small firms of fewer than 100 employees, with medium firms employing 100–499 workers and large firms having more than 500 workers.
- Proportion of small and medium enterprises (SMEs) by industry sector:

Industry Sector	Percentage of SMEs by Sector
Construction	16.3
Professional/Scientific/Technical services	14.6
Retail	11
Accommodation/Food producers	7.8
Transport and Warehouse facilities	7.1
Manufacturing	5.3
Multiple (such as health care, waste management, and entertainment)	20.6

UK AND EU

- Definitions from www.indeed.com refer to small firms as fewer than 500 employees, including sole traders and partnerships of 2+ people
- Towergate Insurance (www.towergateInsurance.co.uk) defines SMEs in a similar way to EU countries, although the small firm may be included in the definition of an SME as up to 250 employees:

Business Size	Number of Employees	Annual Revenue/Turnover
Micro Business	Fewer than 10 employees	Less than £2 million
Small Firm	10–50 employees	£3–£9 million
Medium Business	50–250 employees	£10–£50 million

APPENDIX 2

Site location: _____

Date of assessment: _____

Name(s) of assessor(s): _____

Stage of progress	Main activities	Equipment and processes used	People involved	H&S issues	Security issues
Stage 1: Arrival					
Stage 2: Process					
Stage 3: Process					
Stage 4: Completion or finishing					
Stage 5: Onward movement					

APPENDIX 3

Management actions taken:	Yes (tick)	In part	Complete by	Review date	Review by
a) Established priorities: - Noted "Review" where controls are adequate - Identified high-risk factors with Priority Rating 4–5 - Identified medium-risk factors with Priority Rating 3 - Identified low-risk factors with Priority Rating 1–2					
b) Prepared Plan of Action, with steps needed and time scales set for completion, for: - Priority Rating 4–5 - Priority Rating 3 - Priority Rating 1–2					
c) Established appropriate data bases and recording systems					
d) Provided staff with relevant and sufficient information to be able to complete tasks					
e) Established appropriate consultation procedures with workers					

Management actions taken:	Yes (tick)	In part	Complete by	Review date	Review by
f) Allocated and confirmed parameters of individual responsibility and authority					
g) Identified one or more "competent person(s)" for health and safety risks					
h) Arranged methods for keeping up to date with legislative changes and their impact on the firm					
i) Prepared a full range of policy statements, including health and safety					
j) Notified everyone of details of these policy statements, and confirmed commitment at the most senior level of the firm					

Index

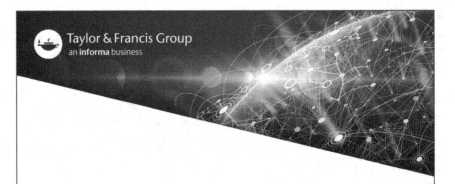

Printed in the United States
by Baker & Taylor Publisher Services